Joint Polish-German Crystallographic Meeting

Zeitschrift für Kristallographie Supplement

2020 Number 40

Joint Polish-German Crystallographic Meeting

February 24–27, 2020,
Wrocław, Poland

Abstracts

DE GRUYTER

ISBN 978-3-11-069287-7
e-ISBN (PDF) 978-3-11-069291-4
e-ISBN (EPUB) 978-3-11-069288-4
ISSN 0930-486X

Library of Congress Control Number: 2020931735

Bibliographic information published by the Deutsche Nationalbibliothek
The Deutsche Nationalbibliothek lists this publication in the Deutsche Nationalbibliografie; detailed bibliographic data are available on the Internet at http://dnb.dnb.de.

Printing and binding: Frantz X. Stückle Druck und Verlag e.K., Ettenheim

www.degruyter.com

Table of content

Inorganic crystal structures I

S01-1
Fundamental Bonding Concepts of Inorganic Chemistry Revisited

S. Grabowsky[1], M. Fugel[2]
[1]University of Bern, Department of Chemistry and Biochemistry, Bern, Switzerland
[2]University Bremen, Bremen, Germany

On the one hand, there are many bonding concepts such as covalency, ionicity, polarity, resonance, hypervalency, hyperconjugation, hypercoordination, dative bonding and more that are of fundamental importance in inorganic chemistry, but that are also debatable. On the other hand, there are numerous bonding descriptors from the regimes of real space, orbital space and energy space such as natural bond orbitals, quantum theory of atoms in molecules or energy decomposition analysis that are useful, but potentially in conflict with each other. Hence, chemical bonding analysis is always an ambiguous enterprise. In this work, we attempt to combine such chemical bonding descriptors into a toolbox set called *complementary bonding analysis*, derive these complementary descriptors not from theory alone, but bring them into the experimentally accessible field of crystallography, and revisit such fundamental bonding concepts mentioned above.

The first example deals with the relationship between resonance, hyperconjugation, hypervalency and bond polarity in the iconic and fundamental oxyanions nitrate, phosphate, sulfate and perchlorate. We measured high-resolution low-temperature single-crystal X-ray diffraction data sets of potassium nitrate, struvite (magnesium ammonium phosphate), lithium sulfate and potassium perchlorate at synchrotrons and home sources. With the help of X-ray wavefunction refinement (XWR), we derived and analyzed complementary bonding descriptors from the regimes mentioned above for this series of anions.

The second example is based on dative N...Si interactions in peri-substituted naphthalene compounds. These dative interactions inflict a certain degree of hypercoordination at the silicon atom which influences the Lewis resonance forms and the valence electron count at the silicon atom. Again, high-resolution low-temperature single-crystal X-ray diffraction data sets were measured at synchrotrons and home sources and analyzed using XWR.

S01-2
Structural properties of whitlockite related synthetic materials: Some recent results

W. Paszkowicz[1]
[1]Institute of Physics PAS, Warsaw, Poland

Whitlockite mineral, Ca9(MgFe)(PO4)6PO3OH and its extraterrestrial form, Ca9NaMg(PO4)7, have close structural relation with Ca3(XO4)2, X = P, V or As. Whitlockite-related materials form a large family. All of them crystallize in R3c S.G. (cell size a~11 Å, c~38 Å). At Ca sites, they accommodate up to ~10% of dopants. Natural minerals contain Na, Mg, Fe and Y, whereas the synthetic ones include various iso- or aliovalent ions. The doped structures can be ordered (e.g. for Mg dopant) or disordered (e.g. RE dopants). Synthetic variants are described e.g. as Ca10M1+(XO4)7, Ca9.5M2+(XO4)7, Ca9M3+(XO4)7, The formulae differ due to the charge-balance condition. Whitlockites are a minor component of living organisms (in scaffold and teeth) and are found in microcalcifications connected with human diseases. The opportunity of doping enables tuning the physical properties.

In this study, a tentative systematics of synthetic whitlockite-related orthophosphates and orthovanadates is presented. Moreover, results of some recent studies on structural and elastic properties of such crystals are presented.

The synthetic crystals from whitlockite family, Ca9.5M(VO4)7 and, Ca9M(VO4)7, were studied, using the X-ray powder diffraction method at ambient conditions, at high temperatures and at high pressures. Laboratory and synchrotron beams were used.

The study presents the whitlockite family, including the recently reported compounds. The experiments resulted in determination of the structural details of a number of vanadate whitlockites. The variation of the structure with temperature, pressure and ion size is analyzed. The coefficient of thermal expansion and bulk moduli of selected materials are determined.

The study extends the knowledge on the structure of whitlockites and provides information on elastic properties of several vanadium-based whitlockites. Such studies are helpful in determining the stability of whitlockites and allow for designing the future application.

S01-3
Novel boron-rich compound $B_{314.6}H_9$

C. Eisele[1], C. B. Huebschle[1,2], S. Mondal[3], S. Dey[4], C. Paulmann[5], S. van Smaalen[1]
[1]University of Bayreuth, Laboratory of Crystallography, Bayreuth, Germany
[2]Bruker AXS GmbH, Karlsruhe, Germany
[3]CSIR-Central Glass & Ceramic Research Institute, Functional Materials & Devices Division, Kolkata, India
[4]RWTH Aachen University, Institute of Crystallography, Aachen, Germany
[5]University of Hamburg, Mineralogisch-Petrographisches Institut, Hamburg, Germany

Since elemental boron polymorphs are difficult to purify, numerous boron-rich compounds arisen from accommodation of even very low quantities of impurity atoms have been discovered and characterised [1],[2]. In particular, the polymorph β-boron tends to form boron-rich compounds by incorporation of impurity atoms into the large vacancies in between its complex framework.

According to Slack *et al.* [3], β-boron crystallises in the centrosymmetric rhombohedral space group *R*-3*m* and has approximately 320.6 boron atoms per hexagonal unit cell which has a volume of V = 2465 Å^3. A well-ordered set of atoms B(1)-B(15), arranged in B_{12} icosahedral clusters and a B_{28}-B-B_{28} cluster of condensed B_{12} icosahedra makes up its framework [3]. The crystal structure of β-boron is complemented by five so-called partially occupied sites (POS) at which a boron atom is accommodated in 3.7 to 27.2 % of the unit cells. These POS are situated in large vacancies in between the framework [3]. As mentioned above, impurity atoms or dopants can be accommodated instead [1],[2].

We will report on our discovery of a new boron-rich compound at the attempt to determine an accurate electron density for β-boron via high-resolution X-ray diffraction at 100 K as successfully accomplished for the polymorphs α-boron and γ-boron [4]. Structure refinement and subsequent

Maximum Entropy Method calculations revealed a new structure lacking of the POS in Slack's model for *β*-boron [3].

References

[1] B. Albert and H. Hillebrecht, Angew. Chem. Int. Ed. 48, 8640 (2009)
[2] A.E. Blagov et al., Crystallogr. Rep. 62, 692 (2017).
[3] G.A. Slack et al., J. Solid State Chem. 76, 52 (1988).
[4] S. Mondal et al., Phys. Rev. B 88, 024118 (2013).

S01-4

Systematics of the Allotrope Formation in Elemental Gallium Films

D. Fischer[1], B. Andriyevsky[2], C. Schön[1]
[1]Max-Planck-Institute for Solid State Research, Stuttgart, Germany
[2]Koszalin University of Technology, Koszalin, Poland

Elemental gallium forms unusual allotropes, whose structures all differ distinctly from the close-packing of spheres typical for nearly all elemental metals. At standard pressure, four different crystalline allotropes are known where the α-Ga allotrope is the thermodynamically stable one. Since the other allotropes can only be obtained under special conditions such as spatial confinement or undercooling, the structural relations among the gallium allotropes are not well-explored.

We present here a systematic investigation of the dependence of the crystal structure formation in elemental gallium films on the deposition temperature and the subsequent annealing procedures using *in situ* X-ray powder diffraction and Raman measurements, complemented by *ab initio* calculations. The films were prepared with the femtosecond pulsed-laser-deposition technique on a substrate kept at constant temperature in the range of −190 °C to 25 °C, followed by cooling+heating cycles in the same temperature range.

Besides preparing α-, β-, γ- and amorphous Ga films as a single phase, a new gallium allotrope, β''-Ga, was synthesized, which is a distorted derivative of the β-Ga crystal structure. Furthermore, no direct transition between the α-Ga phase on one side and the β-, β''-, and γ-Ga phases on the other side was observed in the solid state. Surprisingly, at room temperature amorphous gallium and below −60 °C the α-Ga allotrope, respectively, is formed in the deposited films.

The most surprising discovery is the new β'-Ga allotrope, which we could identify and characterize from the X-ray powder patterns. The existence of this new allotrope was also supported by ab initio calculations of the equations of state of the various Ga-allotropes, which revealed nearly identical energies for α-Ga, β-Ga and β'-Ga.

S01-5

Pyrolysis of *tri*-metaphosphimic acids and its salts

D. Günther[1], C. Kalischer[1], O. Oeckler[1]
[1]Leipzig University, IMKM, Leipzig, Germany

Pyrolysis of *tri*-metaphosphimic acid and its salts leads to the formation of several compounds with unknown compositions. Among these, some known oxonitridophosphates are formed at high temperatures. The pyrolysis of $Na_3(PO_2NH)_3{\cdot}4H_2O$ at 600 °C, for example, yields $NaPO_2NH$.[1]

Tri-metaphosphimic acid can be synthesized by hydrolysis of $(PNCl_2)_3$. The sodium and potassium salts can be obtained by the reaction of $(PNCl_2)_3$ and sodium or potassium acetate, respectively.[2] As $H_3(PO_2NH)_3$ reacts with water, the synthesis of the free acid is best achieved by the protonation of $K_3(PO_2NH)_3$ using $HClO_4$ (w = 60%).[3] In the new modification of $H_3(PO_2NH)_3{\cdot}2H_2O$ discovered in this study (space group $P2_1/c$; a = 7.059(2) Å, b = 13.960(3), c = 11.294(3) Å, β = 124.59(2)), cyclic molecules of *tri*-metaphosphimic acid are stacked along [001]. In contrast to the modification described by Attig (Fig. left),[3] which is characterized by an hexagonal arrangement of coinroll-like stacked molecules, these rods are distorted within the *a-b* plane and are shifted by (0,0,½*c*) with respect to each other (Fig. right). Additionally, in both structures water molecules are located between the stacks, dominantly interacting by hydrogen bonds.

Other metaphosphimates can be obtained by neutralization or ion exchange reactions.[4] Further investigations by temperature-dependent PXRD revealed information about the decomposition of *tri*-metaphosphimates. Starting from e.g. $Na_3(PO_2NH)_3{\cdot}4H_2O$, the solvent water evaporates in two steps up to 200 °C. Between 250 °C and 550 °C, the data indicate the formation of phases like $Na_3P_3O_9$, $Na_5P_3O_{10}$ or related oxonitridophosphates. In contrast, temperatures above 500 °C lead to the formation of $NaPO_2NH$.

[1] N. Stock, Dissertation, Universität Bayreuth, 1998.
[2] M. L. Nielsen, T. J. Morrow, *Inorg. Synth.*1960, *6*, 97.
[3] R. Attig, D. Mootz, *Z. Anorg. Allg. Chem.* 1976, *419*, 139.
[4] N. Stock, E. Irran, W. Schnick, *Z. Anorg. Allg. Chem.* 1999, *625*, 555.

Figure 1

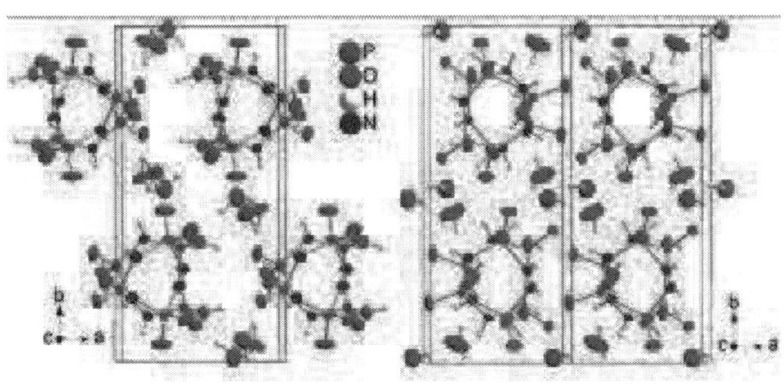

S01-6

CRYSTALLOGRAPHIC STUDIES ON FOSSILIZED ELEPHANT TEETH

N. Białas[1], O. Prymak[1], N. P. Singh[2], R. Patnaik[2], K. Loza[1], M. Epple[1]
[1]University of Duisburg-Essen, Inorganic Chemistry, Essen, Germany
[2]Panjab University, Department of Geology, Chandigarh, India

Introduction

Teeth play a fundamental role in the life of most vertebrates [1]. The Siwaliks (foothills of the Himalayas) are known for the abundance of vertebrate fossils, including proboscideans (elephants) [2]. Fossilized teeth provide information about extinct animals as well as about the past climate and environment [3].

Objectives

Our aim was to perform chemical, crystallographic, and microscopic analyses of fossilized enamel samples, and to study age-related changes, including those of diagenetic origin.

Materials and methods

We studied fossilized teeth of 8 different proboscideans and a tooth of a recent elephant as reference. All fossils were found in the Siwaliks (India), and covered the time range of 0.05 to 14 Ma. The composition of the enamel was determined by elemental analysis, AAS, and UV spectroscopy. The content of water, organic substances and carbonated apatite in the enamel was determined by thermogravimetry. XRD including Rietveld refinement was used to identify the mineral phase (apatite) and to determine its crystallite size. The microstructure and local distribution of elements in the enamel were analyzed by SEM and EDX, respectively.

Results

The calcium content in the fossils was twice as high as in a recent sample, whereas no significant differences in the phosphate content were found. An age-dependency was observed in the content of F, Na and Mg. The apatite crystals were approximately 30-40 nm in size with B-type substitution (CO_3^{2-} for PO_4^{3-}).

Conclusion

Fossilized enamel is a good model to study paleoenvironmental changes. The increasing content of F in the enamel (as function of time) indicated a diagenetic transformation of hydroxyapatite into fluoroapatite.

References

[1] Lübke A., *et al.* 2015. *RSC Adv*, 5, 61612.
[2] Sankhyan A., Chavasseau O. 2018. *Palaeontol Electron*, DOI: 10.26879/844.
[3] Kohn M.J., *et al.* 1999. *Geochim Cosmochim Acta*, 63, 2737-2747.

Micro- and nanocrystalline materials

S02-1
Structure and Morphology of $MoNi_4/MoO_2$@Ni Electrocatalytic Systems for Fast Water Dissociation

E. Zschech[1,2], E. Topal[1,2], Z. Liao[1], J. Gluch[1], M. Löffler[2], S. Werner[3], P. Guttmann[3], G. Schneider[3], J. Zhang[4], X. Feng[4]
[1]Fraunhofer Institute for Ceramic Technologies and Systems, Dresden, Germany
[2]Technische Universität Dresden, Dresden Center for Nanoanalysis, Dresden, Germany
[3]Helmholtz-Zentrum Berlin für Materialien und Energie GmbH, Berlin, Germany
[4]Technische Universität Dresden, Faculty of Chemistry and Food Chemistry, Dresden, Germany

High–performance and low–cost non–precious metal catalysts are of critical importance for electrochemical energy conversion. For a systematic materials selection and for design and synthesis of respective devices, it is essential to clarify the general relationship between physicochemical properties of the materials and the electrocatalytic performance and stability of the system. The design of highly performant and durable 3D electrocatalytic systems requires an optimized hierarchical morphology and surface structures with high activity. The objective of this study is to determine the 3D morphology of hierarchically structured electrocatalytic systems based on multi–scale X–ray computed tomography (XCT) and to characterize the crystalline structure of electrocatalyst nanoparticles using transmission electron microscopy (TEM). We report a novel transition–metal-based materials system: $MoNi_4$ electrocatalysts anchored on MoO2 cuboids aligned on Ni foam ($MoNi_4/MoO_2$@Ni). The high electrocatalytic efficiency of Mo–Ni–based alloy ($MoxNi_y$) electrocatalysts for the hydrogen evolution reaction (HER) is based on their advantageous crystalline structure and chemical bonding. High-resolution TEM images and selected-area electron diffraction patterns are used to determine the crystalline structures of MoO2 and MoNi4. Multiscale XCT provides 3D information on the hierarchical morphology of the $MoNi_4/MoO_2$@Ni composite system nondestructively: Micro-XCT images clearly resolve the Ni foam and the attached needle-like MoO_2 micro cuboids. Laboratory nano-XCT shows that the MoO_2 micro cuboids with a rectangular cross-section of 0.5 x 1 μm^2 and a lengths of 10 to 20 µm are vertically arranged on the Ni foam. $MoNi_4$ nanoparticles with a size of 20 to 100 nm, positioned on single MoO_2 cuboids, were imaged using synchrotron radiation nano-XCT. The application of machine learning algorithms significantly improved the reconstruction quality of the acquired data.

S02-2
In situ observations of single grain behavior during plastic deformation in polycrystalline Ni using energy dispersive Laue diffraction

U. Pietsch[1], M. Shokr[1], L. Strüder[2], C. Kirchlechner[3], C. Genzel[4]
[1]University of Siegen, Physics, Siegen, Germany
[2]pnSensor GmbH, Munich, Germany
[3]Max-Planck-Institut für Eisenforschung GmbH, Düsseldorf, Germany
[4]Helmholtz-Zentrum Berlin für Materialien und Energie GmbH, Berlin, Germany

Energy dispersive X-ray Laue diffraction is applied to investigate the deformation behavior of individual grains in a polycrystalline nickel wire under tensile loading. 38 Laue spots are identified in the Laue pattern which originate from 9 separate grains. The simultaneous measurements of the Laue spot`s position and energy obtained by using a 2D energy dispersive detector, allows to track the evolution of the 9 grains through multiple stages of deformation. Angular and spectral elongation (streaking) of the Laue spots increases as tensile loading is increased and is attributed to macroscopic texture changed and strain due to defect accumulation. On the single grain level, a correlation between crystallographic orientation and strain is investigated. Moreover, spatially resolved anisotropic deformation within a single grain is measured to increase at the grain boundaries. Comparison of the grain specific responses allow for development of a deformation scenario for the whole specimen. The presented experiment demonstrates an alternative protocol for the investigation of deformation mechanisms in polycrystalline materials.

Figure 1

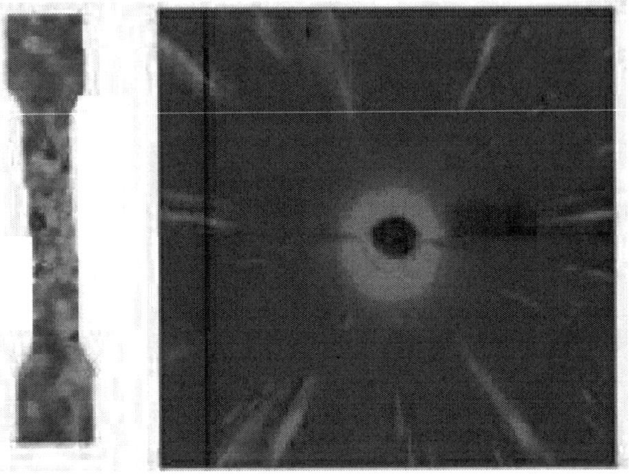

S02-3
Microsecond-resolved look at the very early stages of quantum dot formation

A. Magerl[1]
[1]Friedrich-Alexander-University Erlangen-Nürnberg, Erlangen, Germany

To access the very early stages of nucleation and growth of nanoparticles in liquid media, we have developed a novel setup, where the reactants pushed by high power syringe pumps through a Y-shaped micromixer are immediately liberated as a free jet. With synchrotron X-ray scattering as a probe, this provides access to chemical reaction times as short as 10 µs. Diffraction data with CdS as a prototype example for quantum dot formation show a three-step pathway. Between 10 and 2500 µs, the CdS quantum dot formation starts with a rapid formation of primary clusters driven by the fast diffusion of cadmium and sulphur ions. Further particle growth is by cluster attachment. At this early stage, the particles are not yet crystalline. This reaction pathway is supported by ab initio theoretical calculations.

S02-4

Diffusion and segregation kinetics in immiscible metallic nanoalloys Au-Pt

I. Smirnov[1], Z. Kaszkur[1]
[1]Institute of Physical Chemistry, Polish Academy of Sciences, Warsaw, Poland

Au-Pt system is considered nonmiscible but Au-Pt nanoalloys (NA) can be prepared and were extensively studied. There are however no structural studies in-situ following transport phenomena in various chemical environments. The macroscopic phase diagram is not fully applicable. The NA above some temperature approach equilibrium and separate into Au-rich and Pt-rich phases. Within the miscibility gap several different configurations are possible: core-shell [1], onion and Janus structures [2].

Nowadays, the NA ordering is mainly studied by computational simulations [3]. Computations however suffer several limitations: (a) elementary processes are not fully known [4]; (b) potential parameters are based on a limited experimental basis; (c) the studied models are quite small.

We base our research on the in-situ XRD. Varying conditions of experiment, different diffusion mechanisms can be separately triggered. As we showed [4] CO chemisorption depletes nanocrystal of vacancies and the diffusion (slower) has to follow different routes. Any rearrangements of atoms cause changes in lattice parameters, which can be detected by XRD method. Passing from one (quasi) stable state to another, we explore influence and path of transport phenomena.

We have synthesized Au-Pt@SiO2 NA. The thermally induced phase separation and transport phenomena were monitored over a few weeks using in-situ XRD. It encompassed cyclic surface segregation of Au in He atmosphere followed by segregation of Pt triggered by chemisorption of CO. The obtained and analyzed peak evolution proves that both phases remain next to each other within the same nanocrystal and we can deduce the mass transport between them. Kinetics of the processes allows insight into the diffusion rate. The observed Pt segregation in CO atmosphere is visibly slower than Au segregation in He.

[1] J. Mater. Chem., 2011,21, 4012
[2] J. Phys. Chem. C 2010, 114, 11026;
[3] J. Phys. Chem. C 2012, 116, 19, 10814;
[4] PCCP., 2015, 17, 28250.

S02-5

Application of electron diffraction tomography on incommensurate crystal structures

E. Götz[1], U. Kolb[1,2], S. Plana Ruiz[1], H. J. Kleebe[1], M. Trapp[1]
[1]TU Darmstadt, Applied Geoscience, Darmstadt, Germany
[2]Johannes Gutenberg Universität, Mainz, Germany

Introduction

Incommensurate crystal structures often show twinning, chemical ordering or small crystalline domains. Automated Diffraction Tomography (ADT) allows to use 3D electron diffraction data from such nano-sized crystals for single crystal structure analysis. (Rozhdestvenskaya et al., 2011; Kolb, 2012)

Objectives

Using Labradorite as an example the applicability of ADT on strongly twinned and modulated structures was tested.

Materials and methods

Labradorite is part of the plagioclase series stable at 50-70% anorthite content showing ~100nm wide Ca/Na-rich lamellae. The special play of colour, the labradorescence, results from the interference of the light reflected by the lamellae. (Jin & Xu, 2017) The crystals exhibit strong twinning at different magnitudes and an incommensurately modulated crystal structure with two modulation vectors. Transmission electron microscopy (TecnaiF30, 300keV, STEM for imaging and crystal tracking, Gatan CCD Ultrascan 4000 for diffraction patterns), ADT data acquisition in nano electron diffraction mode (10μm aperture for 50nm beam diameter), data processing (eADT and PETS) and structure solution (SIR2014 and JANA) was used.

Results: Based on ADT data from a single 100nm lamella the mean structure was solved. The lamella size correlates with the observed labradorescent colour and the anorthite content, accordingly. Subsequently, a full refinement based on the detected modulation vectors was performed.

Conclusion

ADT turns out to be a suitable method for solving incommensurately modulated structures. The results obtained for Labradorite are in good agreement with the structure solution by X-ray diffraction.

References

Jin S., Xu H. 2017: American Min., vol.102, 1328-1339
Kolb U. 2012: Automated Diffraction Tomography. Uniting Electron Crystallography and Powder Diffraction. Springer, 314-326
Rozhdestvenskaya E., Mugnaioli E., Czank M., Depmeier W., Kolb U., Merlino S. 2011: Mineral. Magazine, vol.75, No.6, 2833-2846

S02-6

On the Investigation of the adsorbate evolution in mesoporous silicon by combining Anomalous Small-Angle X-ray Scattering and physisorption of Xenon

A. Hoell[1], E. Gericke[1,2], D. Wallacher[1], G. Greco[1], M. Krumrey[3], K. Rademann[2], S. Raoux[1,4]
[1]Helmholtz-Zentrum Berlin für Materialien und Energie GmbH, Institute for Nanospectroscopy, Berlin, Germany
[2]Humboldt Universität zu Berlin, Institute of Chemistry, Berlin, Germany
[3]Physikalisch Technische Bundesanstalt, Berlin, Germany
[4]Humboldt Universität zu Berlin, Institute of Physics, Berlin, Germany

Mesoporous materials are increasingly used in energy related applications as catalysts or thermoelectric materials. Methods to characterize the pore structures are physisorption and small-angle scattering. Gas physisorption in mesoporous materials and the associated capillary hysteresis are largely exploited for the characterization of porous solids. To date, the major hurdle lies in a reliable description of the state of the confined fluid, which is usually given by measuring a macroscopic observable, i.e. the amount of adsorbed gas.

Despite computational methods, in situ techniques combining gas physisorption with X-ray scattering methods showed to be valuable tools to get deeper insights into gas adsorption phenomena.

Combining the different contrasts of SAXS and SANS and contrast matching, a more detailed, locally resolved description of the process can be given by the analysis of the scattering signals of the material pore structure. However, clear assessment of the adsorption process was still missing since the adsorbate evolution in the mesoporous host could be only indirectly investigated.

We present a novel analytical approach, which combines SAXS, ASAXS, XANES and gas physisorption in a single experiment, enabling a direct characterization of the adsorbate phase. The noble gas Xenon (Xe) was used for the physisorption that was done at the Xe boiling point (T = 165 K) in a mesoporous silicon membrane with around 9 nm wide pores.

The instrumental setup used allowed us to reach the Xenon L3 X-ray absorption edge at 4.781 keV. The combination of the three experiments, ASAXS, XANES and physisorption were done on different points of the adsorption and desorption branch of the isotherm. Thus, from the resonant scattering curves of Xe the mesoscopic evolution of the adsorbate (multilayer formation, capillary condensation and desorption) could be directly investigated.

The experimental setup for ASAXS in combination with physisorption will be described in more detail.

Solid state physics in crystallography

S03-1

Random structure search: Solving the kesterite-stannite puzzle in $(Cu,Ag)_2ZnSnSe_4$ solid solution

D. Fritsch[1], S. Schorr[1,2]
[1]Helmholtz-Zentrum Berlin für Materialien und Energie GmbH, Structure and Dynamics of Energy Materials, Berlin, Germany
[2]Freie Universität Berlin, Institute of Geological Sciences, Berlin, Germany

$Cu_2ZnSnSe_4$ and $Ag_2ZnSnSe_4$ are both crystallising in the kesterite structure with the structurally similar stannite structure being energetically slightly less favourable. In the solid solution, however, there is experimental evidence that for some intermediate concentrations the stannite structure is energetically favoured. The reason for that structural change is so far not completely understood.

Here, we're using first-principles calculations based on density functional theory to shed some light into the structure-property relations in the $(Cu,Ag)_2ZnSnSe_4$ solid solution. In order to simulate the different concentrations within the solid solution, we're employing the supercell approach based on the respective end members in the kesterite and stannite structure. The Ag and Cu cations are distributed randomly within the supercell, thereby creating several structure models for the solid solutions for further analysis. All random structure models are geometry optimised employing different exchange and correlation functionals, namely the PBEsol and the recently developed SCAN functionals. In order to obtain more reliable electronic and optical properties, selected optimised structures are subjected to one-shot calculations employing the more accurate hybrid functional HSE06.

This work made use of computational resources provided by the North-German Supercomputing Alliance (HLRN), and the Curta and Dirac HPC facilities of the Freie Universität Berlin and the Helmholtz-Zentrum Berlin, respectively.

S03-2

Interactions of Ruddlesden-Popper Phases and Migration-Induced Field-Stabilized Polar Phase in Strontium Titanate

C. Ludt[1], M. Zschornak[1]
[1]Technische Universität Bergakademie Freiberg, Freiberg, Germany

Modification of ideal crystal structures are accompanied by changes of the electronic structure which may lead to new interesting properties and applications.

Ruddlesden – Popper (RP) phases in strontium titanate [1] are well known. Recently another phase has been found by Hanzig et al. [2]. The so-called migration-induced field-stabilized polar (MFP) phase builds during electroformation and exhibits promising properties, i.e. pyro- and piezoelectricity.

Here we present the electronic structures of the homologous series of RP phases $SrO(SrTiO3)_n$ modeled by means of density functional theory (DFT) including the compounds $n = 0 \ldots 3, \infty$, as well as an evaluation of possible interaction of RP phases with the MFP phase accompanied with a comparison to experimental data [3]. The DFT modeling is based on supercells with embedded RP phases in bulk $SrTiO_3$. The RP phases distort the surrounding cubic $SrTiO_3$ unit cells. In the resulting tetragonal unit cells the ions move on the picometer scale. In summary, the calculations suggest that RP phases may be involved in the formation process of the MFP phase by inducing strain on the surrounding SrTiO3.

[1] S.N. Ruddlesden and P. Popper, *The compound $Sr_3Ti_2O_7$ and its structure,* Acta Crystallographica 11, 54–55 (1958).
[2] J. Hanzig, M. Zschornak, F. Hanzig, E. Mehner, H. Stöcker, B. Abendroth, C. Röder, A. Talkenberger,G. Schreiber, D. Rafaja, S. Gemming, D.C. Meyer, *Migration-induced field-stabilized polar phase in strontium titanate single crystals at room temperature,* Physical Review B 88, 024104 (2013).
[3] C. Richter, M. Zschornak, D. Novikov, E. Mehner, M. Nentwich, J. Hanzig, S. Gorfman, D.C. Meyer, *Picometer polar displacements in strontium titanate determined by a new approach of resonant x-ray diffraction,* Nature Communications 9, 178 (2018).

S03-3

Thermal diffuse scattering x-ray measurements on a metal-organic framework compound and on thiourea with high energy photons

J. Büscher[1], M. Stękiel[1], D. Spahr[1], E. Haussühl[1], O. Ivashko[2], M. von Zimmermann[2], A. C. Dippel[2], B. Winkler[1]
[1]Goethe University Frankfurt, Institute of Geoscience, Frankfurt/Main, Germany
[2]Deutsches Elektronen Synchrotron (DESY), Hamburg, Germany

Thermal diffuse scattering (TDS) can be used to obtain elastic stiffness coefficients [1]. Often encountered problems arise from surface scattering or radiation damage. Here, we explore the use of 100 keV photons to collect TDS data to overcome these problems.

We measured TDS of the metal-organic framework (MOF) Zn-guanidinium formate, [C(NH2)3]Zn[(HCOO)3], and of thiourea, SC(NH2)2. Zn-guanidinium formate (ZnGF) belongs to perovskite-like MOFs that are known to display many fascinating properties, such as multiferroicity [2]. Thiourea is a well investigated ferroelectric that exhibits a variety of pressure- [3] and temperature-dependent phases [4].

At the PETRA III P21.1 beamline at DESY, we carried out experiments using a photon energy of 100 keV. The samples had dimensions of about 200 μm × 50 μm × 50 μm (ZnGF) and 500 μm × 1000 μm × 2000 μm (thiourea) and were of excellent quality. We measured both compounds at different temperatures (235 to 295 K) but without overstepping phase boundaries. There was no radiation damage even after prolonged beam exposure. We are currently analyzing the data and evaluating the results; the method will be discussed at the DGK.

The authors gratefully acknowledge the Deutsche Forschungsgemeinschaft (DFG, HA 5137-5, Wi 1232) and the BMBF (05K16RFA, 05K16RFB) for financial support of this study. We acknowledge DESY (Hamburg, Germany), a member of the Helmholtz Association, for the provision of experimental facilities. Parts of this research were carried out at PETRA III, beamline P21.1.

Fig. 1: TDS of thiourea at 235 K in the (001) plane. Left: experimental data obtained with a PILATUS CdTe detector. Right: results from DFT-based model calculation.

[1] B. Wehinger et al., PRL 118, 035502 (2017)
[2] P. Jain et al., J. Am. Chem. Soc. 131, 13625-13627 (2009)
[3] A. Banerji and S. K. Deb, J. Phys. chem. B, 37, 10915-10919 (2007)
[4] G. J. Goldsmith and J. G. White, J. Chem. Phys., 31, No. 5 (1959)

Figure 1

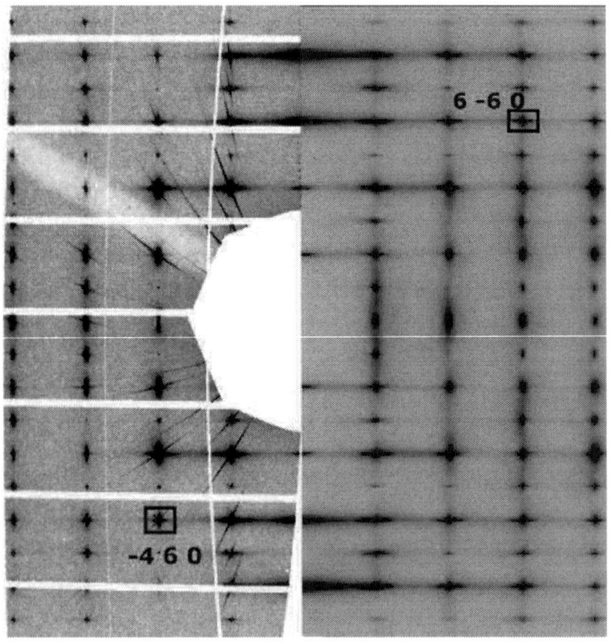

S03-4
Verification of the de Wolff hypothesis concerning the symmetry of β-MnO_2

P. Fabrykiewicz[1], R. Przeniosło[1], I. Sosnowska[1], F. Fauth[2], D. Oleszak[3]
[1]University of Warsaw, Faculty of Physics, Institute of Experimental Physics, Warsaw, Poland
[2]CELLS-ALBA, Cerdanyola del Valles, Spain
[3]University of Warsaw, Faculty of Materials Science and Engineering, Warsaw, Poland

Introduction
Pyrolusite, β-MnO_2, has a tetragonal rutile-type ($P4_2/mnm$) crystal structure [1]. Below the Néel temperature T_N=92 K β-MnO_2 shows an antiferromagnetic ordering superimposed on a helical modulation of the Mn^{4+} magnetic moments [2], however de Wolff proposed an orthorhombic structure of β-MnO_2 [3].

Objectives
The goal of this study was to describe the known modulated magnetic ordering of β-MnO_2 [2,4] with the magnetic superspace groups formalism [5, 6] and to verify de Wolff hypothesis [3].

Materials and methods
Synchrotron radiation (SR) diffraction measurements were performed at the powder diffraction beamline BL-MSPD of the ALBA synchrotron [7]. Neutron powder diffraction studies were performed previously using the flat-cone diffractometer E2 at the reactor BER-II [4].

Results
Both SR and neutron powder diffraction data support the orthorombic rather than the tetragonal symmetry [7]. The magnetic superspace groups formalism does not allow for a helical ordering in β-MnO_2 but the experimental results can be modeled with a spin density wave ordering within the orthorhombic magnetic superspace group Pnnm.1'(00γ)s00s [7].

Conclusions
The findings described above were recently published [7] and they can motivate a revisit of the models describing the magnetic properties of β-MnO_2.

Bibliography
[1] Baur, W.H. (2007). *Crystallogr. Rev.,* 13, 65-113.
[2] Yoshimori, A (1959). *J. Phys. Soc. Jpn,* 14, 807-821.
[3] Wolff, P.M. de (1959). *Acta Cryst.,* 12, 341-345.
[4] Regulski, M., Przeniosło, R., Sosnowska, I., & Hoffmann, J.-U. (2003). *Phys. Rev. B,* 68, 172401.
[5] Wolff, P.M. de, Janssen, T. & Janner, A. (1981). *Acta Cryst A,* 37, 625-636.
[6] Perez-Mato, J.M., Ribeiro, J.L., Petříček, V. & Aroyo, M.I. (2012). *J. Phys. Condens. Matter,* 24, 163201.
[7] Fabrykiewicz P., Przeniosło R., Sosnowska I., Fauth F. & Oleszak D. (2019). *Acta Cryst A,* 75, 889-901.

Bio-Crystallography I: Signalling, macromolecular interactions and other new structures

S04-2

Structural and functional principles of a novel family of nucleic acid helicases

J. J. Roske[1], S. Liu[1], B. Loll[1], U. Neu[2], M. C. Wahl[1,3]
[1]Freie Universität Berlin, Laboratory of Structural Biochemistry, Berlin, Germany
[2]Freie Universität Berlin, Laboratory for Biochemistry of Viruses, Berlin, Germany
[3]Helmholtz-Zentrum Berlin für Materialien und Energie GmbH, Macromolecular Crystallography, Berlin, Germany

Nucleic acid-dependent NTPases (ndNTPases) are motor proteins that can utilize the energy from NTP hydrolysis to remodel DNA, RNA or nucleic acid-protein complexes. Based on these activities, these enzymes are essential players in DNA replication and repair, genome maintenance and virtually all aspects of gene expression and regulation in all three domains of life. Based on the presence of conserved sequence motifs, the domain composition, the oligomeric state and the NTP and nucleic acid preferences, ndNTPases have been grouped into several superfamilies (SFs) and families. Many of these enzymes can unwind nucleic acid duplexes, *i.e.* they can act as nucleic acid helicases. The molecular mechanisms underlying helicase activities can differ between members of different SFs and families. Many ndNTPases that act as helicases can translocate along their nucleic acid substrate in an NTP-driven fashion while other act *via* cycles of binding, deforming and releasing their substrates. We have studied a so far underexplored class of superfamily II helicases. Members of this family are widely distributed in bacteria, archaea, plants and fungi and exhibit a unique set of helicase-associated domains. We characterized the nucleic acid binding and unwinding activities of one member of this family *in vitro* and determined crystal structures of the protein in isolation and bound to a nucleic acid substrate. Our results delineate new elements in this helicase family that are important for nucleic acid binding and unwinding.

S04-3

F2X Universal and F2X Entry – chemically highly diverse libraries for crystallographic fragment screening at HZB

J. Wollenhaupt[1,2], T. Barthel[1], A. Metz[2], G. Lima[3], D. Wallacher[4], T. Hauss[1], M. Gerlach[1], C. Feiler[1], M. C. Wahl[1,5], U. Müller[3], G. Klebe[2], M. S. Weiss[1]
[1]Helmholtz-Zentrum Berlin für Materialien und Energie GmbH, Macromolecular Crystallography, Berlin, Germany
[2]Philipps-Universität Marburg, Drug Design Group, Marburg, Germany
[3]Max IV Laboratory, BioMax, Lund, Sweden
[4]Helmholtz-Zentrum Berlin für Materialien und Energie GmbH, Sample Environments, Berlin, Germany
[5]Freie Universität Berlin, structural biochemistry, Berlin, Germany

Being strongly established in industry, crystallographic fragment screening gains more and more popularity in academia as well. Fragments are individually soaked in protein crystals and binding is detected via X-ray diffraction experiments. The resulting protein-fragment structures provide starting points for development of larger and more potent binders.

Crystallographic fragment screening experiments and their analysis can be performed within a seamless workflow at HZB in a highly efficient manner. We provide improved methods of sample-handling, robot-assisted beamlines as well as highly automated processing of the experimental data up to the identification of even weakly bound fragments. However, the outcome of the workflow depends to a large extend on the quality of the fragment library chosen. Higher count and chemical diversity of the resulting fragment hits increases the chances for successful design of follow-up compounds.

In collaboration with the Drug Design Group in Marburg we developed the F2X Universal Library as well as its sub-selection the F2X Entry Screen. Both are optimized to be maximally diverse and thus for high coverage of the chemical space. Here, we present the validation of the F2X Entry Screen using the model system Endothiapepsin as well as real-life cases. The data was analyzed using the FragMAX project manager webapp that is currently being developed at the BioMAX group at MAX IV. The results show high hit rates, significantly outscoring previous screens on the same targets. Additionally, our setup of the libraries as ready-to-use 96-well plates with dried-in compounds, facilitates fragment screening also without DMSO, which we confirmed experimentally.

Figure 1

S04-4

Engineered variants of β-lactoglobulin with multiple binding sites for ligands

J. Loch[1], J. Sławek[1], J. Barciszewski[2], P. Wróbel[1], W. Minor[3], K. Lewiński[1]
[1]Jagiellonian University, Faculty of Chemistry, Kraków, Poland
[2]Polish Academy of Sciences, Center for Biocrystallographic Research, Institute of Bioorganic Chemistry, Poznań, Poland
[3]University of Virginia, Department of Molecular Physiology and Biological Physics, Charlottesville, Virginia, United States

Bovine β-lactoglobulin (BLG) belongs to the lipocalin family. All lipocalins share a conserved structural motif, made by antiparallel β-barrel, which is also primary binding site for ligands. In physiological pH natural and recombinant lactoglobulin exists as homodimer with molecular weight about 36 kDa.

In the course of our studies on the properties of lactoglobulin mutants, we have created a library of BLG variants possessing mutations inside binding pocket. The aim of our studies was to systematically screen library of BLG mutants for affinity to selected low molecular weight ligands to identify mutants with special binding properties.

All BLG mutants used for the structural studies were expressed in *E.coli* and purified according to published protocol [1]. Protein-ligand complexes were crystallized by hanging drop technique. X-ray diffraction data were collected at BESSY Berlin (beamline 14.1), APS Argonne (beamline 19 ID) and at XtaLAB Synergy (Rigaku).

Structural analysis showed that typically, BLG can accommodate a single ligand molecule in the β-barrel [2]. However, it was found that some mutants, especially those possessing mutations in positions 39, 56, 105 and 107 are able to bind several drug molecules per protein dimer (Fig. 1). In these structures ligands are bound not only in the β-barrel but also at additional sites located at dimer interface and at the entrance to the β-barrel.

It seems that unusual binding properties of some mutants originate from the type of mutations present in the biding pocket, type of ligand, pH and ionic strength of crystallization solution. Currently, our efforts are focused on precise identification of the conditions that are necessary to activate non-standard binding sites in lactoglobulin molecule.

References

[1] Mol Biotechnol 58 (2016) 605-618.
[2] Int J Biol Macromol 114 (2018) 85-96.

Fig. 1. Crystal structure of lactoglobulin mutant with four desipramine molecules bound per dimer.

Figure 1

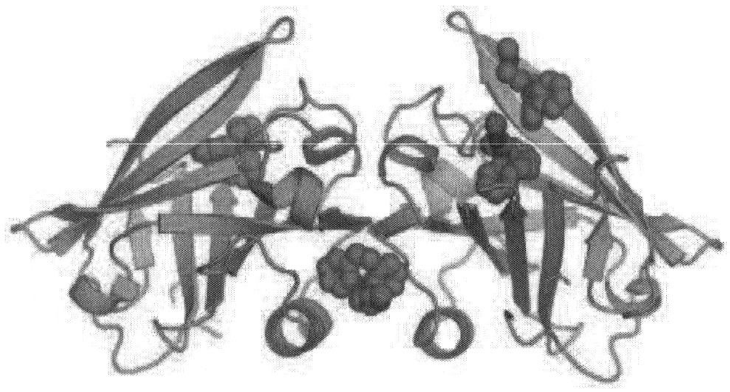

S04-5

Crystal structures of β-D-galactosidase from *Arthrobacter* sp. 32cB - cold adaptation and active site architecture

G. Bujacz[1], M. Rutkiewicz[1], A. Bujacz[1]
[1]Łódź University of Technology, Institute of Molecular and Industrial Biotechnology, Łódź, Poland

The investigated cold-adapted β-D-galactosidase (EC 3.2.1.23) from *Arthrobacter sp. 32cB* (*Arth*βDG) is a large, five-domain enzyme containing 1010 amino acids, with the catalytic domain in a form of typical TIM-barrel. This enzyme is active as a dimer and exhibits both; high hydrolytic and transglycosylation activity.

A mesophilic βDGs are widely used in food industry to hydrolyze lactose. However, replacing it with a cold-adapted enzyme such as *Arth*βDG, would lead to a number of advantages.

Crystal structures of *Arth*βDG have been determined in an unliganded form resulting from diffraction experiments conducted at 100 K and at room temperature. A detailed comparison of those two structures of the investigated enzyme was performed in order to estimate differences in their molecular flexibility and rigidity and to study structural rationalization for the cold-adaptation.

Obtained crystal structures of *Arth*βDG complexes with: galactose, IPTG, ONPG and lactose enabled characterization of the active site and determination of the residues that take part in substrate biding.

The research was supported by grant 2016/21/B/ST5/00555 from the NCN coordinate by AB.

References

Rutkiewicz-Krotewicz M., Pietrzyk-Brzezinska A., Wanarska M., Cieslinski H., Bujacz A. Crystals, 2018, 8, i1

Rutkiewicz, M.; Bujacz, A.; Bujacz, G. *Biochim. Biophys. Acta - Proteins Proteom.* 2019, 1867, 776–786.

Rutkiewicz-Krotewicz, M. Bujacz, A.; Wanarska, M.; Wierzbicka-Wos, A.; Cieslinski. *Int. J. Mol. Sci*, 2019, 20, i17.

S04-6

The rough nanotexture of calcium carbonate biocrystals is not an indication of growth by particle attachment, but by crystallization from an amorphous precursor

A. Checa[1,2], E. Macías-Sánchez[3], A. Sánchez-Navas[4], N. Lagos[5]
[1]Instituto Andaluz de Ciencias de la Tierra, CSIC-Universidad de Granada, Armilla, Spain
[2]University of Granada, Stratigraphy and Paleontology (Faculty of Sciences), Granada, Spain
[3]Max Planck Institute of Colloids and Interfaces, Potsdam, Germany
[4]University of Granada, Mineralogy and Petrology, Granada, Spain
[5]University of Santo Tomás, Centro de Investigación e Innovación para el Cambio Climático, Santiago de Chile, Chile

There is presently a consensus that the growth of biocrystal proceeds by non-classical crystallization, whereby particles of amorphous calcium carbonate (ACC) adhere to the growing crystals to form a solid or liquid phase, which later transforms into crystalline. Whereas the existence of precursor ACC has been confirmed in a number of cases, the only evidence for particle attachment (PA) is the so-called nanogranular texture displayed by most biocrystals, characterized by both surface and internal roughness, with nanogranules being interpreted as the aggregation units. Detailed SEM and SEM-EBSD analysis of aragonite produced by molluscs (nacre) and granular calcite of barnacles reveals that nanoprotrusions consistently align and/or elongate along the <100> direction in nacre, and along the <-441> directions in calcite. In both cases, these directions are the strongest periodic bond chains (PBCs). Based on the notion that both biominerals crystallize from a precursor ACC and on the present knowledge on their nanostructure, we hypothesize that during crystallization into either aragonite or calcite, organic molecules present in the ACC phase are expelled from the corresponding PBCs to other directions. The expelled organic molecules tend to accumulate in the surroundings of the continuous elongate crystalline cores formed along the PBCs, where they stabilize small amounts of ACC, thus forming the amorphous pellicles usually recognized. Accordingly, the nanogranularity results from the transformation of the amorphous into the crystalline phase, and not from the alleged PA process. What is interpreted as a nanogranular surface is, in fact, the highly irregular crystallization front.

Structure-property-relationships I

S05-1
MIDAS - Mapping the Irregularity of Distribution of Atoms in Space

W. Hornfeck[1]
[1]Institute of Physics of the Czech Academy of Sciences, Prague, Czech Republic

Several thousand of structural descriptors for structure-property/structure-activity relationships are currently known and used in molecular chemoinformatics. When it comes to similar structural descriptors for solid state structures, however, almost nothing seems to be known. We will show how uniform distribution measures might fill this gap in allowing the quantification of the (ir)regularity of atoms in space, based on the knowledge of the atomic coordinates of a crystal structure, and how this might be used in establishing novel structure-property relationships. For this purpose we present the conceptual outlook for a dedicated software solution, MIDAS, currently under development.

S05-2
How different should similar diffraction patterns be to indicate polymorphs?

M. Główka [1]
[1]Łódź University of Technology, Department of Chemistry, Łódź, Poland

Introduction
The possibility of patenting new polymorphs of the approved API results in increasing interest on the subject in the pharmaceutical industry. Powder X-ray diffraction is the fastest, very simple and univocal method used in polymorphs identification just by comparing respective diffraction patterns.

Objectives
In view of the above another question arises, namely, how (much) different the crystal structures of the two samples should be to call them polymorphs? Several examples of suspicious "polymorphs" taken from patents as well as substances claimed wrongly as the same polymorphs only on the basis of their X-ray diffraction patterns will be presented.

Materials and methods
Theoretically, the similarity of powder diffraction patterns indicates the similarity of the compared substance. In practice, experimental conditions, habit and quality of crystals may influence diffraction patterns. To study the common crystal disturbances affecting diffraction, simulated diffraction patterns from selected single crystal structures from CSD were used.

Results and conclusions
- It is not possible to correlate changes in similar diffraction patterns with a specific structural features or preparation and measurement conditions if crystal structures are not known;
- There is no general rule or figure, which allows objectively to indicate true polymorphs only on the basis of similar diffraction patterns;
- In the case of unknown crystal structures being compared, claiming the discovery of a new polymorph based only on the differences in diffraction patterns needs to be supported by distinctly different physical properties;
- The practical solution of the problem in United States Pharmacopeia: *"..... identity of diffraction patterns is established if the scattering angles of the 10 strongest reflections agree to within ±0.20 degrees and if the relative intensities of these reflections do not vary by more than 20%."*

S05-3
Neutron single crystal diffraction investigation of tetragonal compounds in the $Cs_2CuCl_{4-x}Br_x$ mixed system

N. van Well[1,2], B. Pedersen[3,4], S. Heuss-Aßbichler[5], A. Schönleber[2], S. van Smaalen[2]
[1]Ludwig-Maximilians-University Munich, Department for Earth- and Environmental Sciences, Section for Crystallography, Munich, Germany
[2]University of Bayreuth, Laboratory of Crystallography, Bayreuth, Germany
[3]Technical University Munich, Heinz Maier-Leibnitz Zentrum (MLZ), Garching, Germany
[4]Ludwig-Maximilians-University Munich, Munich, Germany
[5]Ludwig-Maximilians-University Munich, Department of Earth and Environmental Sciences, Section for Mineralogy, Petrology and Geochemistry, Munich, Germany

The $Cs_2CuCl_{4-x}Brx$ mixed system is rich in structural and magnetic phases, which can be separated by their tetrahedral or octahedral Cu^{2+} environment. For the octahedral Cu^{2+} environment of the $Cs_2CuCl_{4-x}Br_x$ mixed system, the compounds are typical quasi 2-D antiferromagnets [1]. For the structure investigation of the tetragonal compounds of this mixed system, synchrotron powder diffraction was used. The structure analysis down to 4K for $Cs_2CuCl_{2.2}Br_{1.8}$ shows no phase transition and the tetragonal symmetry *I*4/mmm is retained over the whole temperature range [2]. However, new single-crystal neutron diffraction experiments indicate a very small orthorhombic distortion at low temperature. The structure solution shows a subgroup relationship for the investigated composition of this mixed system. The realisation of the new tetragonal phase of composition Cs_2CuCl_4 is possible using specific crystal growth conditions below 281K [2]. The susceptibility measurements of Cs_2CuCl_4 show similar magnetic behaviour like the magnetic susceptibility and the magnetisation curves of the tetragonal compounds $Cs_2CuCl_{2.9}Br_{1.1}$, $Cs_2CuCl_{2.5}Br_{1.5}$ and $Cs_2CuCl_{2.2}Br_{1.8}$, and present consistent results for typical quasi 2-D antiferromagnets [2]. Several magnetic reflections corresponding to the propagation vector $k = (0, 0, 0)$ are observed for this compound with neutron diffraction experiments below the magnetic phase transition at $T_N = 11.3K$ confirming its antiferromagnetic nature.

[1] P. T. Cong, B. Wolf, N. van Well, A. A. Haghighirad, F. Ritter, W. Assmus, C. Krellner, M. Lang, IEEE Transactions on Magnetics 50, 2700204 (2014)

[2] N. van Well, C. Eisele, S. Ramakrishnan, T. Shang, M. Medarde, A. Cervellino, M. Skoulatos, R. Georgii, S. van Smaalen, Cryst. Growth Des. 2019, doi.org/10.1021/acs.cgd.9b01035

S05-4

Old tricks for a new dog: Crystallographic toolboxes for halide perovskites

J. Breternitz[1], S. Schorr[1,2]
[1]Helmholtz-Zentrum Berlin für Materialien und Energie GmbH, Structure and Dynamics of Energy Materials, Berlin, Germany
[2]Freie Universität Berlin, Institute of Geological Sciences, Berlin, Germany

Hybrid halide perovskites have made a quite spectacular appearance in the field of photovoltaics, not only because device efficiency has shot over 25 % within 10 years of their development,[1] but also because of their specific behaviour that is yet to be fully understood. In structural terms, these materials compare to the oxide perovskites ABO_3 in many respects[2] but also hold some features that are rather distinct and largely related to the molecular cation occupying the A-cation site.[3]

Therefore, a structural categorisation of these materials is greatly beneficial to understand the underlying principles of structure and property relationships. At the core of this work, we will present a fairly comprehensive group-subgroup relationship applied to halide perovskites and double perovskites deriving from the cubic perovskite aristotype in the form of a Bärnighausen tree.[4] This is seconded with a discussion of the different distortion modes applying to halide perovskites: atom shifts, octahedral tilting and A-cation orientation, with the latter being a distinct mechanism in hybrid halide perovskites. Furthermore, we will elucidate the implications for the properties and phase transitions given the specific space group settings of the different crystal structures. Finally, we will extend the discussion on the use of solid-state chemical approaches, such as the bond-valence-sum concept, for the understanding of halide perovskites and the abstraction of their properties.

Figure 1: group subgroup relationship between the different phases of $MAPbI_3$. The octahedral tiltings are given in Glazer notation.[5]

References

[1] *https://www.nrel.gov/pv/cell-efficiency.html*, accessed 28/10/2019.
[2] J. Breternitz, S. Schorr, *Adv. Energy Mater.* **2018**, *8*, 1802366.
[3] J. Breternitz, F. Lehmann, S. A. Barnett, H. Nowell, S. Schorr, *Angew. Chem.* doi: 10.1002/ange.201910599
[4] J. Breternitz, S. Schorr, in preparation.
[5] A. M. Glazer, *Acta Crystallogr. A* **1972**, *28*, 3384.

Figure 1

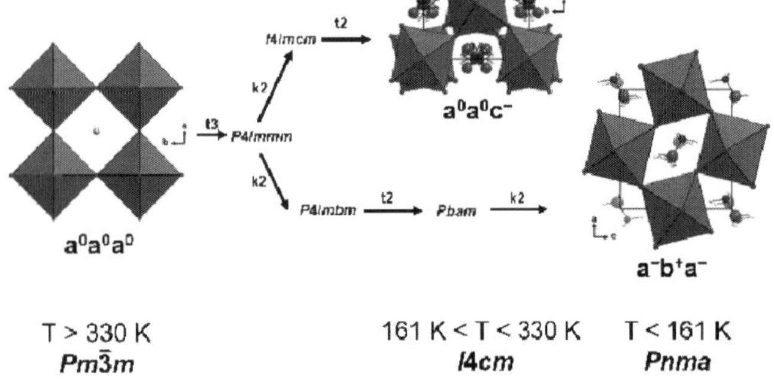

S05-5

Raman Spectroscopy – a quantitative method for analyzing stress states in quartzite

K. J. Huenger[1], M. Danneberg[1,2], J. Acker[1,2,3], S. Herold[1,2,3]
[1]Brandenburg University of Technology Cottbus-Senftenberg, Research & Materials Testing Institute , Cottbus, Germany
[2]Brandenburg University of Technology Cottbus-Senftenberg, Chair Building Materials&Chemistry, Cottbus, Germany
[3]Brandenburg University of Technology Cottbus-Senftenberg, Chair Physical Cemistry, Senftenberg, Germany

Introduction

Alkali silica reaction is one of the worldwide most investigated damaging reactions in concrete. An important part to understand it is the SiO2 release from natural grains. It is well-known, that the solubility is mainly surface controlled. Therefore, it is necessary to get more information on the surface properties of aggregates, especially the stress state of quartz crystallites. A common method is the light polarization microscopy at thin sections. Using this method the stressed areas can be visualized, but not evaluated quantitatively.

Objective

To find a method to quantify the stress state on quartzite surfaces for each quartz grain, a combination of polarization microscopy (for visualize the stressed zones) and Raman spectroscopy (to quantify them) was used. So maybe, it could be possible to evaluate the stress states in different locations of aggregate grains.

Results

Fig.1 shows a Raman spectrum of an unstressed grain area (on the left). A different spectrum (on the right) with surprising new signals at 1600 and 1300 cm-1 was found in regions with expected stress. The micrograph in **Fig. 2** demonstrates deformed quartz structures in a contact zone between quartz crystallites. And exactly there, the signals at 1600 cm-1 can be measured by Raman spectroscopy. Additionally the intensity of well crystallized quartz areas measured at 466 cm-1 is reduced accordingly there.

Further investigations show that such signals cannot be measured in all areas of quartz grains with coloured shift. Other grains have different orientations and show not such signals at 1600 cm-1.

Conclusion

The reasons for this effect are not clear yet. But it can be concluded, maybe that it is possible to use the found signals at 1600 cm-1 of RAMAN spectroscopy to evaluate such stressed areas quantitatively. A better differentiation and quantification of real stressed areas can be used for a better understanding of dissolution processes of quartz grains in natural aggregates.

Figure 1

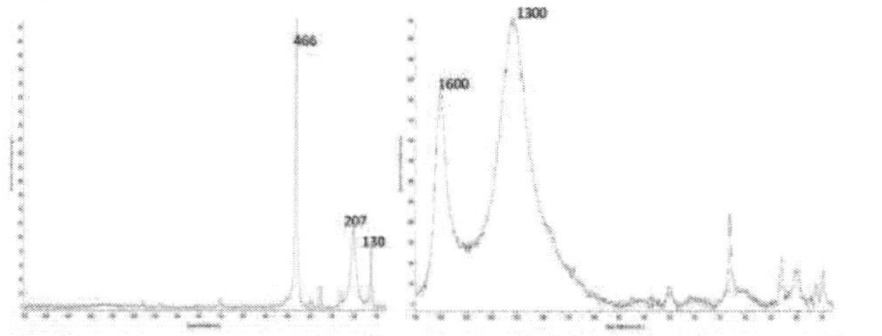

Figure 1- left: Raman spectrum of an unstressed area with typical quartz signals at 466, 207 and 130cm-1; right: Raman spectrum of areas with expected stress, new signals are measurable at 1600 and 1200cm-1

Figure 2

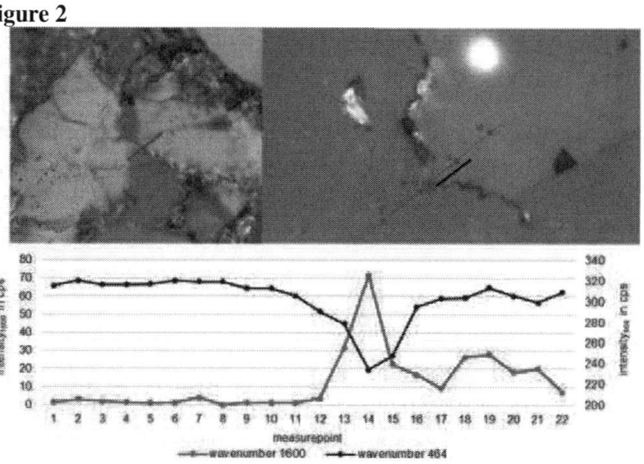

Figure 2 The upper left side shows the polarized light microscopy with the position of the Raman measurement. The Dark field view in the upper right corner shows the line scan mode with the Raman Microscope. The black line shows the range where the 1600 cm^{-1} signal increases. The development of the intensity of the Raman Signals with 464 cm^{-1} and 1600 cm^{-1} is shown below.

S05-6

Structural control of thermomechanical properties of piezoelectric rare-earth calcium oxoborates

M. Münchhalfen[1], J. Schreuer[1], E. Mehner[2], H. Stöcker[2], C. Reuther[3], J. Götze[3]
[1]Ruhr University Bochum, Institute of Geology, Mineralogy and Geophysics , Bochum, Germany
[2]Technische Universität Bergakademie Freiberg, Institute of Experimental Physics, Freiberg, Germany
[3]Technische Universität Bergakademie Freiberg, Institute of Mineralogie, Freiberg, Germany

Monoclinic rare-earthcalcium oxoborates $R\mathrm{Ca_4O(BO_3)_3}$ *(RCOB)* recently gathered interest as candidates for high-temperature piezoelectric sensing applications since they combine favorable properties like high melting point at around 1770 K, no reported structural phase transitions, high piezoelectric sensitivity and high electric resistivity. Since the *RCOB* structure offers different possibilities for cation substitution, tuning of physical properties is principally possible.

Large single crystals of *RCOB* (*R* = Er, Y, Dy, Gd, Sm, Nd, La) were grown from melt using the Czochralski method. Their structural properties were studied at ambient conditions on as cast and on annealed and subsequently quenched samples using X-ray diffraction methods. Dilatometry and resonant ultrasound spectroscopy were employed to investigated their thermo- and electromechanical properties between 100 K and 1473 K.

Substitution of rare earth cations with larger size leads to a decrease in the bulk modulus, whereas the maximum value of the piezoelectric effect increases. Contrary to the reported lack of phase transitions, the investigated physical properties undergo reproducible discontinuities between 900 K and 1300 K which are characteristic for a glass-like transition. Structural investigations indicate a correlation with dynamic disorder of *R* and Ca on the independent cation sites at higher temperatures. The transition temperature as well as the specific type of disorder are both related to size of the rare earth ion.

Acknowledgments: The authors gratefully acknowledge financial support of the DFG (PAK921/1, SCHR761/4: Structure/property relationships and structural instabilities of high-temperature piezoelectrics of the oxoborate family).

in situ/in operando studies

S06-1

In Operando Diffraction Radiography and Tomography on Li-Ion Batteries

A. Schökel[1], A. Senyshyn[2], V. Baran[2]
[1]Deutsches Elektronen Synchrotron (DESY), Hamburg, Germany
[2]Technical University Munich, FRM II, Garching, Germany

Diffraction radiography and tomography were used to study different commercial Li-ion batteries. Depending on the data processing either 'direct' parameters, such as phase distributions, or more abstract, derived parameters, like degree of lithiation of the graphite anode or even local state of charge, can be extracted. Especially the lithium distribution is an interesting parameter and is surprisingly inhomogeneous throughout the volume of a commercial battery cell.

Figure 1 Gas pressure cell (left) and cross section of the diffraction plane (right), measurements in mm

References

[1] H. Auer et al., *Inorg. Chem.* 2017, 56, 1072–1079
[2] A. Götze et al., *Phys. B* (Amsterdam, Neth.) 2018, 551, 395-400
[3] T. B. Massalski et al., 1990, 2, 2051

Figure 1

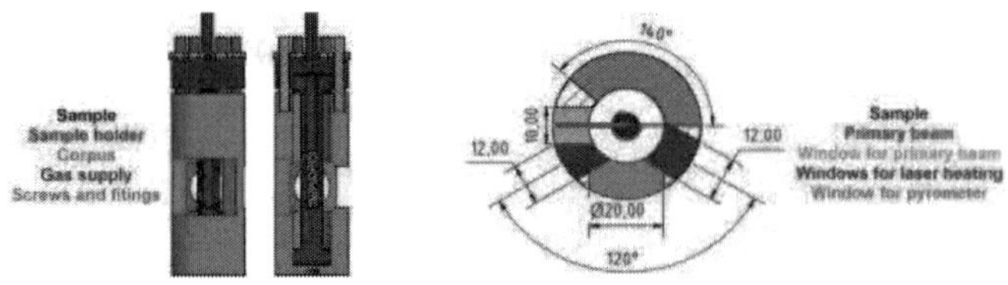

S06-2

A new type of sapphire single-crystal gas pressure cells for *in situ* neutron scattering

R. Finger[1], H. Kohlmann[1]
[1]Leipzig University, Inorganic Chemistry, Leipzig, Germany

Introduction

In recent years, science and industry invested in new technologies as an alternative for fossil fuels. Therefore, solid gas reactions are getting high attention. In order to investigate hydrogenation reactions and pathways of intermetallic compounds, a sapphire single-crystal gas pressure cell for elastic neutron scattering has been validated and is used since 2013 at D20 (Institut Laue-Langevin, Grenoble, France). Using a single crystal, all Bragg reflections from the sample holder can be avoided, resulting in high quality neutron diffraction data [1,2].

Objectives

The cell design is aiming for minimized mechanical tension and maximised data quality. Therefore, a new corpus has been developed. Besides, the sample holder has been slightly changed, having less contribution to the diffraction pattern.

Materials

The sample holder is a leuco-sapphire single crystal. It is surrounded by a corpus of stainless steel with neutron absorbing varnish. Windows for neutrons, heating and temperature surveillance are given. The heating system is a 2x100 W diode laser, the temperate is surveyed by a pyrometer. Conditions up to 100 bar H_2 and 700 K have been tested successfully.

Results

The developed gas pressure cell has been validated via the deuteration of Pd powder. At a constant temperature of 430(1) K, injections of deuterium gas showed the composition and decomposition of palladium deuterides, in accordance with the phase diagram [3].

Conclusion

A new gas pressure cell to follow *in situ* hydrogenation reactions of intermetallic compounds has been developed and is validated via the deuteration of Pd powder. Its maximal operating conditions are 100 bar of H_2/D_2 and 700 K.

S06-3

Remarkable anisotropic thermal expansion in coordination compounds

G. Gallo[1,2], S. Bette[1], R. E. Dinnebier[1], K. Yadava[3], Z. Chen[3], J. J. Vittal[3]
[1]Max-Planck-Institute for Solid State Research, Stuttgart, Germany
[2]University of Salerno, Department of Chemistry and Biology "A. Zambelli", Salerno, Italy
[3]National University of Singapore, Department of Chemistry, Singapore, Singapore

Introduction

Coordination compounds consist of a metallic centre coordinated by molecules or ions. Depending on the ligands and central metal atom or ion, these compounds have interesting features concerning optical, thermal and mechanical properties. In this work, three isotypical complexes showing photosalient effect and a MOF were investigated by X-ray powder diffraction (XRPD). The study of the thermal behaviour revealed a remarkable anisotropic expansion.

Objectives

Variable temperature *in situ* XRPD (VT-XRPD) measurements were performed in order to investigate the dynamic structural changes upon heating.

Results

VT-XRPD data were collected and subjected to series of sequential Rietveld and Pawley refinements. Lattice parameters and volume were extracted and thermal expansion coefficients calculated using the software PASCal. From detailed structural analysis, the mechanism of the anisotropic thermal expansion can be proposed. All these compounds show that the remarkable anisotropic expansion is driven by coordination bonds between the metallic centre and the ligands. Additionally, in the three isotypical complexes, π-π and F···H-C interactions also play a key role.

Conclusions

VT-XRPD was used to obtain quantitative information on the dynamic changes in the structures during the heating process. A deep understanding on the origin mechanism of the anisotropic thermal behaviour is crucial the design of new functional materials with tuneable properties.

Fig. 1. Plots and molecular representation showing anisotropic thermal expansion results of a case study.

Figure 1

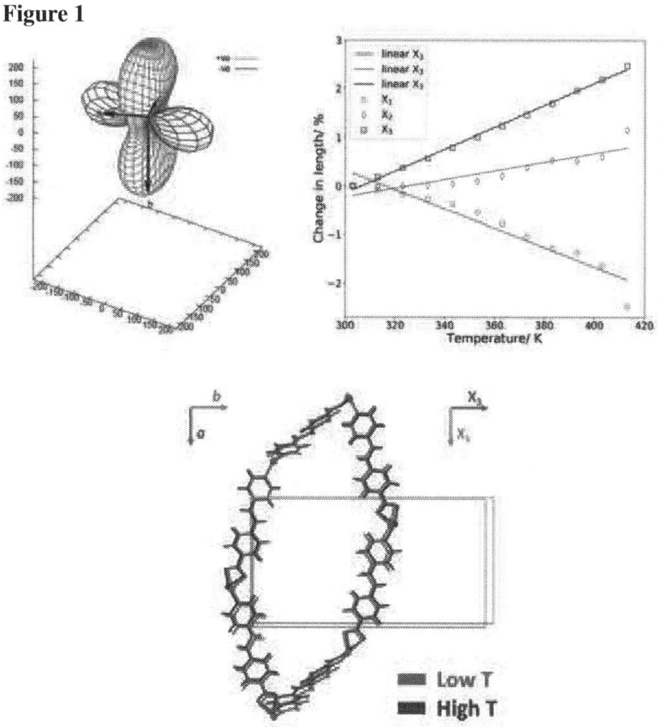

S06-4
Metal-Support Interactions in Gold Quasicrystals Deposited on Cerium (IV) Oxide Catalyst Revealed by In-Operando Powder X-Ray Diffraction Coupled with Mass Spectrometry

M. Zielinski[1], Z. Kaszkur[1]
[1]Institute of Physical Chemistry, Polish Academy of Sciences, Catalysis on Metals, Warsaw, Poland

Our research focused on the *in-situ* dynamic structure changes of the surface and the crystal lattice of heterogeneous catalyst 9.4%$_{wt.}$ Au/CeO_2. These changes were induced by the preferential oxidation of CO running in the presence of H_2 (PROX). Au-decorated ceria was an efficient catalyst reaching over 80% conversion of CO to CO_2 with high selectivity against H_2O production.

The assumed hypothesis suggested that the interaction with adsorbents results in the immediate perturbation of the initially relaxed structure of the catalyst surface.

As the expected crystal lattice changes would be very small, *in-operando* Powder X-Ray Diffraction (PXRD) offering ultimate precision of the lattice constant changes was selected. Coupled Mass Spectrometry (MS) enabled on-line monitoring of the catalyst chemical performance. Structure data analysis was aided by Transmission Electron Microscopy (TEM) imaging.

CeO_2 is an easily reducible material and even more prone to reduction when neighbouring AuNPs. Ceria crystal lattice expanded under the atmosphere of CO and H2 (see Fig. 1). Ce^{3+} cation has larger radius than the Ce^{4+} ion, so the crystal structure expands with the increasing concentration of Ce^{3+} [G. Brauer & K. A. Gingerich, *J. Inorg. Nucl. Chem.*, 16 (1960) 1 87-99] or their redistribution to the CeO_2 nanoparticle core. Pure CeO_2 was inactive in CO oxidation under studied reaction conditions and remained unaffected by the hydrogen atmosphere at ~150°C. Gold Multiple Twinned Particles (MTPs, see Fig. 2) were discovered as their presence noticeably affected the X-Ray diffraction patterns and they were directly observed in TEM micrographs. Au seemed to be crucial for activation of oxygen and formation of CO_2. CeO_2 surface served in the reaction as a large adsorption site for CO and it bound activated oxygen moieties.

Fig. 1. The change of apparent lattice parameter of CeO2 as a function of gas atmosphere composition.

Fig. 2. The HR TEM image of Au MTP – decahedron.

Figure 1

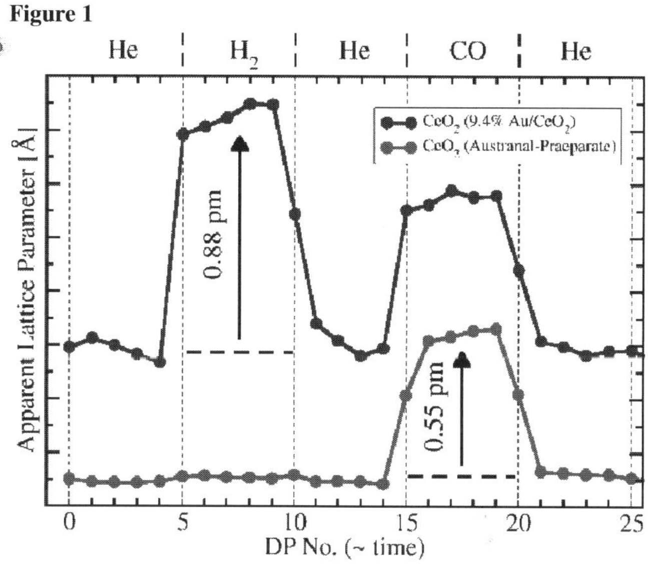

Figure 2

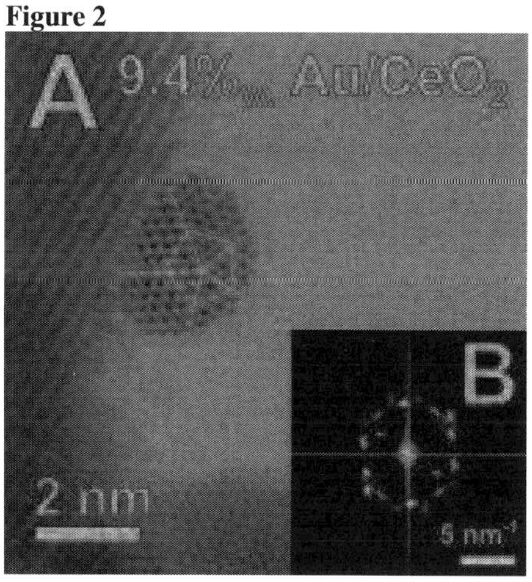

S06-5
In-situ XRD and PDF investigation of $MF_3.3H_2O$ (M = Fe, Cr) in controlled atmosphere: accessing new phases with controlled chemistry

G. Nenert[1], K. Forsberg[2]
[1]Malvern Panalytical, Almelo, Netherlands
[2]KTH Royal Institute of Technology, Stockholm, Sweden

Introduction

Iron fluoride ($FeF_3.nH_2O$) shows high capacity as cathode material for lithium-ion batteries combined to low toxicity and low cost. The water content of iron fluoride has been shown to be of prime importance in the performances of the cathode. So far, the various synthesis route do not allow for a precise water content control, especially on the low amount regime which is the most interesting range of composition. In addition, CrF_3 has been shown to increase significantly the conductivity of LiF film [2]. Consequently, it is of interest to look for the in-situ formation of the various $MF_{3-x}(OH)_x.nH_2O$ phases (M = Cr, Fe).

Objectives

We aimed to investigate in-situ the formation of $MF_{3-x}(OH)_x.nH_2O$ (M = Fe, Cr) phases to control the precise crystal chemistry using a controlled atmosphere environment and look for new phases not reported so far using XRD and PDF analysis.

Materials and Methods

To reach our goals, we have been investigated MF3.3H2O (M = Fe, Cr) powder using in-situ high temperature XRD and PDF analysis taking advantage of the transmission geometry. The use of capillaries enables us to control the generated atmosphere upon heating and thus assessing a rich chemistry.

Results

Some of the results are presented in Figure 1a) about the FeF3-x(OH)x.nH2O phases and in Fig. 1)b for $CrF_{3-x}(OH)_x.nH_2O$ Precise controlled of the water content of the $FeF_{3-x}(OH)x._nH2O$ phase could be reached with n ranging from 1/3 to 0 with about 10 new pure phases. The controlled in-situ decomposition of $CrF_3.3H_2O$ led to the formation of a new $CrF_{3-x}(OH)_x$ pyrochlore which was characterized structurally and magnetically.

Conclusion

We were able to obtain new phases with precise chemistry which have been obtained thank to the self generated atmosphere using in-situ heating with capillaries.

[1] Kim et al., Adv. Mater. 2010, 22, 5260–5264, Ma et al., Energy Environ. Sci. 2012, 5, 8538–8542
[2] Tetsu Oi Materials Research Bulletin, 19, 451-457 (1984)

Figure 1

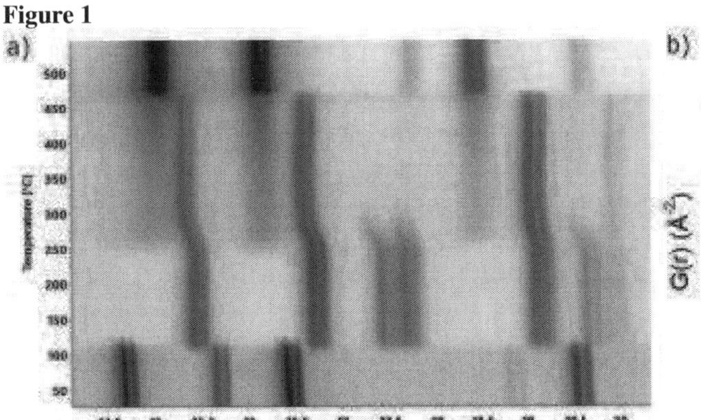

S06-6

Operando XRD studies of selected NaxMnO2 and MoS2 electrode materials for Na-ion batteries.

A. Kulka[1], K. Walczak[1], A. Plewa [1], J. Płotek[1]
[1]AGH University of Science and Technology, Department of Hydrogen Energy, Kraków, Poland

Sodium-ion batteries (SIBs) recently has gained a lot of interests due to the speculation about poor Li supply and its" rising prices resulting from the enormous demand for Li-ion batteries. SIBs are especially considered for stationary applications where their low cost may be the most important criterion. Since then, the hope to see a commercial SIBs in the near future is likely and it pushes the scientists" efforts towards optimization of the electrode materials in order to increase SIB"s performance. Na_xMnO_2 oxides (NMO) and MoS_2 have drawn research attention as a SIB"s positive and negative electrodes respectively. However the exact electrochemical reaction with Na is still unclear for these two groups of materials and needs proper examination. In this report we show operando XRD investigation of NMO and MoS_2 materials upon electrochemical Na (de)sodiation and correlate the structural transition with their performance. The NMO were obtained using high-temperature synthesis while MoS_2 via hydrothermal method. The operando XRD measurements were performed using an in-house made *operando* XRD cell. NMO oxides showed different P2 (P63/mmc) to P"2 (C2/c) phase ratio depending on the synthesis conditions. We found that single P2 phase NMO material shows "smooth" potential profile while in biphasic P2/P"2 Na_xMnO_2 potential curves show a stepwise character linked to Mn^{3+}/Mn^{4+} charge, Na^+ ion /vacancy orderings, and character of phase transition during (de)sodiation at higher potentials. Our findings show that, by tuning the pristine phase of NaxMnO2 materials it is possible to change the (de)sodiation mechanisms from P2+P"2 -type → OP4- type phase (observed for biphasic P2/P"2 Na_xMnO_2) to the intergrowth mechanism (P2 Na_xMnO_2), hence influencing their performance.

The work was supported by Polish National Science Center under grant nr 2016/21/D/ST5/01658.

Fig. 1. Operando XRD results for single P2 Na_xMnO_2

Fig. 2. Operando XRD results for biphasic P2/P"2 Na_xMnO_2

Figure 1

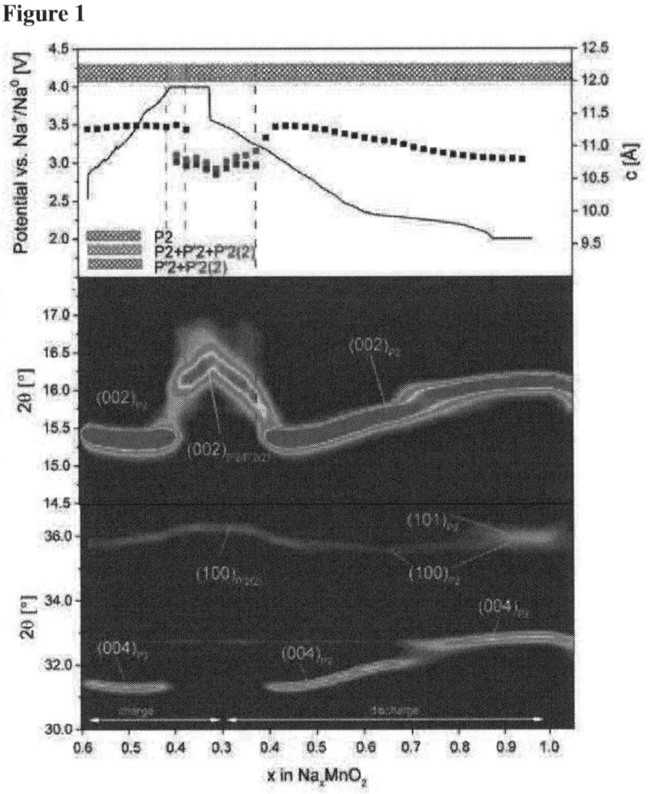

Figure 2

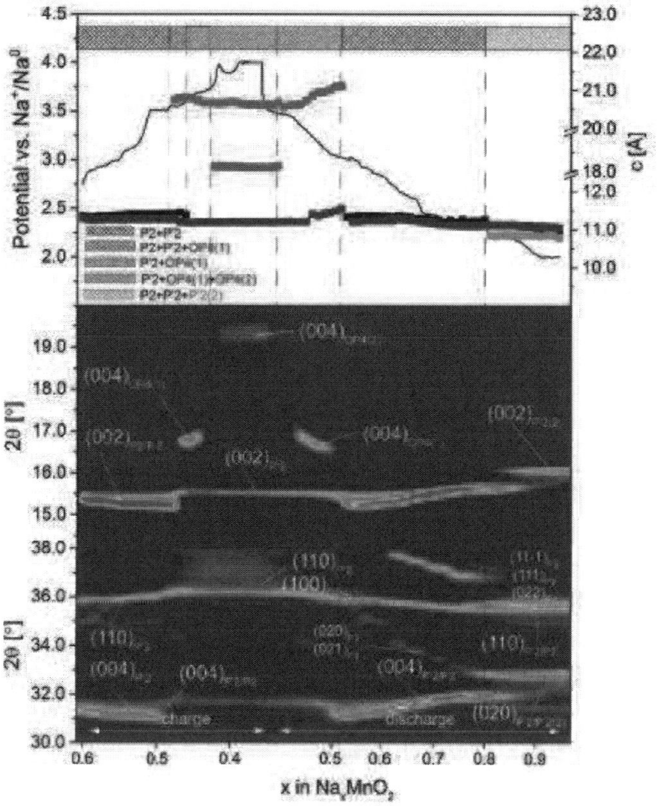
Potential vs. Na^+/Na^0
c [Å]
P2+P2
P2+P2+OP4(1)
P2+OP4(1)
P2+OP4(1)+OP4(2)
P2+P2+P2(2)
2θ [°]
charge
discharge
x in Na_xMnO_2

Bio-Crystallography II: Enzymes

S07-1

Crystal structure and biochemical characterization of the plastic-degrading *Ideonella sakaiensis* MHETase

G. Palm[1], L. Reisky[2], D. Böttcher[2], H. Müller[2], E. Michels[1], M. Walczak[2], L. Berndt[1], M. S. Weiss[3], U. Bornscheuer[2], G. Weber[3,1]
[1]Universität Greifswald, Molecular Structural Biology, Greifswald, Germany
[2]Universität Greifswald, Biotechnology and Enzyme Catalysis, Greifswald, Germany
[3]Helmholtz-Zentrum Berlin für Materialien und Energie GmbH, Macromolecular Crystallography, Berlin, Germany

Despite many advantages, synthetic polymers (plastics) have become a global problem since they are omnipresent and not accessible to environmental degradation processes. One example is the polyester polyethylene terephthalate (PET) which amounts to about 18% of all synthetic polymers ever synthesized. The extreme durability of PET debris has rendered it a long-term environmental burden and at the same time, current recycling efforts still lack sustainability. Two recently discovered bacterial enzymes that specifically degrade PET represent a promising solution for polyester recycling. First, *Ideonella sakaiensis* PETase, a structurally well-characterized consensus α/β-hydrolase fold enzyme, converts PET to mono-(2-hydroxyethyl) terephthalate (MHET). MHETase, the second key enzyme, hydrolyzes MHET to the PET educts terephthalate and ethylene glycol. We have determined the crystal structures of active ligand-free MHETase, MHETase bound to a nonhydrolyzable MHET analog or the chemically related compound benzoate. The structure of MHETase was solved by molecular replacement, employing the MORDA MR-pipeline and with a related feruloyl esterase as selected search model. Based on the structural data, we achieved a rational improvement of MHETase activity and were able to extend its substrate spectrum by mutations as documented by *in vitro* activity assays and differential scanning fluorimetry (DSF). MHETase, which is reminiscent of feruloyl esterases, possesses a classic α/β-hydrolase domain and a lid domain conferring substrate specificity. In the light of structure-based mapping of the active site, activity assays, mutagenesis studies and a first structure-guided alteration of substrate specificity towards bis-(2-hydroxyethyl) terephthalate (BHET) reported here, we anticipate MHETase to be a valuable resource to further advance enzymatic plastic degradation together with PETase.

Figure 1

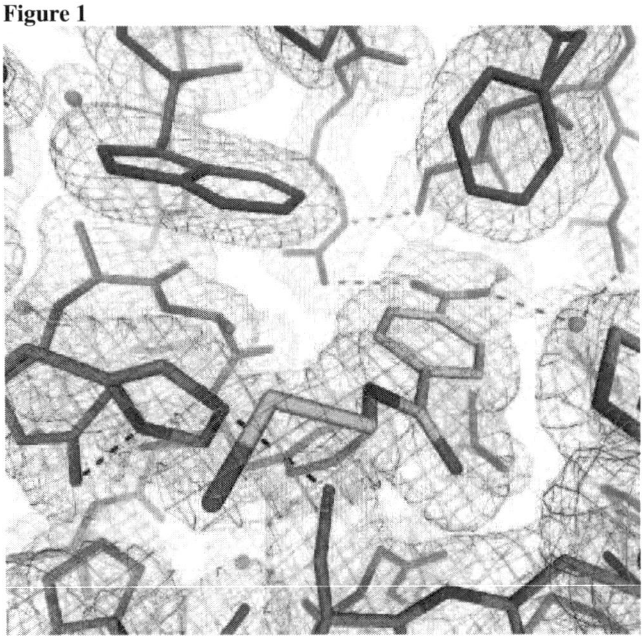

S07-2

Half way to hypusine – structural insights into human deoxyhypusine synthase

P. Grudnik[1], E. Wątor[1], P. Wilk[1]
[1]Jagiellonian University, Malopolska Centre of Biotechnology, Structural Biology Core Facility, Kraków, Poland

Hypusination is a posttranslational modification of lysine to an unusual amino acid, hypusine. In the first step of hypusination, the 4-aminobutyl moiety of spermidine is transferred to the e-amine group of lysine by deoxyhypusine synthase (DHS) forming deoxyhypusine. In humans, hypusine is present only in the eukaryotic translation initiation factor 5A (eIF5A).

eIF5A is involved in translation elongation and it facilitates protein synthesis via resolving polyproline-induced ribosomal stalling. Its role is indispensable in the synthesis of proline repeat-rich proteins. Hypusination of eIF5A is essential for its activity in promoting cell proliferation. Furthermore, there is clear evidence of eIF5A alterations are involved in different diseases, such as diabetes and cancer.

The aim of the presented study was to investigate substrate specificity of DHS using macromolecular crystallography and to assess the impact of newly recognized pathological DHS mutations on protein stability, activity and structure.

DHS and its two mutants were expressed, purified and crystalized. We have solved six high-resolution crystal structures of apo DHS and in complexes with natural substrates. Based on crystal structures and activity tests we showed that despite almost identical binding of spermidine and spermine only spermidine can serve as a proper substrate of deoxyhypusine formation. Furthermore, our data demonstrate that, against the previous studies, no conformational changes occur in DHS structure upon spermidine binding. Therefore, we hypothesize that significant conformational change is needed for the further progress of reaction occurring as a result of eIF5A substrate recognition. Availability of high-quality structural data will aid the design of novel DHS inhibitors for potential clinical applications and can significantly advance our understanding of DHS deficiency syndrome.

Figure 1. Human DHS crystal structure.

Figure 1

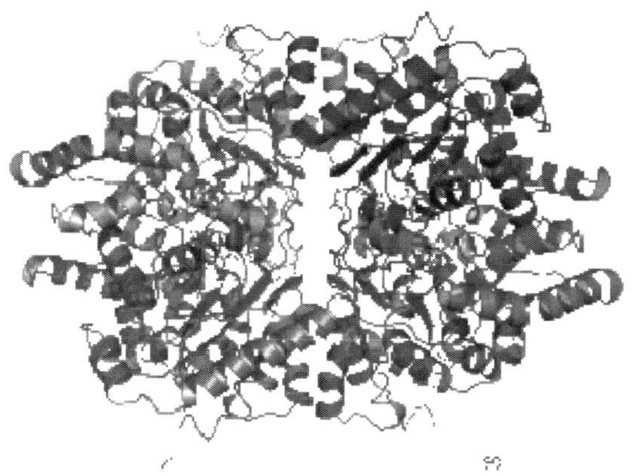

S07-3

Identification and characterization of the bottromycin epimerase BotH expands the catalytic scope of alpha/beta-hydrolases

J. A. Köhnke[1,2]
[1]Helmholtz Institute for Pharmaceutical Research Saarland, Saarbrücken, Germany
[2]Helmholtz Center for Infection Research, Structural Biology of Biosynthetic Enzymes, Saarbrücken, Germany

Ribosomally synthesized and post-translationally modified peptides (RiPPs) are a rapidly growing natural product superfamily with interesting and diverse bioactivities. Bottromycins are RiPPs with potent activity against gram positive bacteria, including the problematic human pathogens MRSA and VRE.[1] The biosynthesis of bottromycins has recently received attention through a series of in vitro experiments, including macroamidine formation and the macroamidine-dependent proteolytic processing of a pathway intermediate.[2-5]

Despite their ribosomal origin, bottromycins contain a *D*-Aspartate (*D*-Asp), which was proposed to arise through the spontaneous epimerization of an *L*-Asp during biosynthesis. We discovered that the highly unusual alpha/beta-hydrolase (ABH) BotH from a bottromycin biosynthetic gene cluster catalyzes the post-translational epimerization of *L*-Asp to *D*-Asp *in vitro*. The combination of biochemical data with the structures of BotH and its complex with product allow us to propose a mechanism for this reaction. Surprisingly, BotH was also able to bind bottromycins with high affinity. BotH-bottromycin complex structures revealed a mode of binding similar to that of substrate which implies an additional function in bottromycin biosynthesis. Overall, our findings expand the catalytic repertoire of ABH enzymes to include epimerization and bioinformatic analyses of BotH homologs suggest that ABH enzymes with this activity may be found in diverse biosynthetic gene clusters.

References

[1] Shimamura, H., et al., *Angew Chem Int Ed Engl* **2009,** *48* (5), 914-7, doi: 10.1002/anie.200804138.
[2] Franz, L., et al., *J Am Chem Soc* **2017,** *139* (50), 18158-18161, doi: 10.1021/jacs.7b09898.
[3] Mann, G., et al., *Chembiochem* **2016,** *17* (23), 2286-2292, doi: 10.1002/cbic.201600406.
[4] Schwalen, C. J., et al., *J Am Chem Soc* **2017,** *139* (50), 18154-18157, doi: 10.1021/jacs.7b09899.
[5] Sikandar, A., et al., *J Am Chem Soc* **2019,** *141* (25), 9748-9752, doi: 10.1021/jacs.8b12231.

S07-4

Structural characteristics of D-2-hydroxyacid dehydrogenase family

J. Kutner[1], D. Matelska[2], I. G. Shabalin[3], K. Ginalski[2], K. Woźniak[1], W. Minor[3]
[1]University of Warsaw, Biological and Chemical Research Centre, Department of Chemistry, Warsaw, Poland
[2]University of Warsaw, Laboratory of Bioinformatics and Systems Biology, Centre of New Technologies, Warsaw, Poland
[3]University of Virginia, Department of Molecular Physiology and Biological Physics, Charlottesville, Virginia, United States

Introduction

D-2-Hydroxyacid dehydrogenases (2HADHs) catalyze the reversible NAD(P)H-dependent stereospecific reduction of 2-ketocarboxylic acids to (*R*)-2-hydroxycarboxylic acids.

Objectives

The long and complex evolution and broad sequence diversity hinder functional annotations for the uncharacterized members of the 2HADH family. We have focused on the structural characteristics of the novel 2HADHs enzymes.

Materials and Methods

We have crystallized novel 2HADHs enzymes coupled with various substrates and cofactors. To characterize the functional divergence between the enzyme subfamilies, we tested the activity of two *Sinorhizobium meliloti* enzymes with 24 ligands and measured their kinetic parameters.

Results

We have revised the classification of the whole 2HADHs family, which comprises 22 subfamilies, out of which 13 new subfamilies are previously unknown. We have shown that GHPR family consists of the three evolutionarily separated subfamilies that we named GHRA, GHRB, and GHRC. We solved crystal structures of various complexes, with the most important one with 2-keto-D-gluconate (PDB ID: 5v7n, Fig. 1) as the bulkiest substrate co-crystallized with a 2HADHs family. Both enzymes are symmetrical homodimers composed of two globular α/β/α domains: coenzyme-binding domain and catalytic domain. We were able to identify pyruvate and 2-keto-D-gluconic acid as highly active and selective substrates for *Sm*GhrA and *Sm*GhrB, respectively.

Conclusions

Systematic analyses of active site environments provided key insights into the residues important for substrate selectivity. Our family-wide sequence and structural comparison (*http://2hadh.bioreproducibility.org*) proved general importance of several active sites thus extending our understanding of its catalytic machinery.

Acknowledgments

The project is supported by Foundation for Polish Science/European Union under the European Regional Development Fund (TEAM TECH CORE FACILITY/2017-3/4, POIR.04.04.00-00-31DF/17-00)

Figure 1

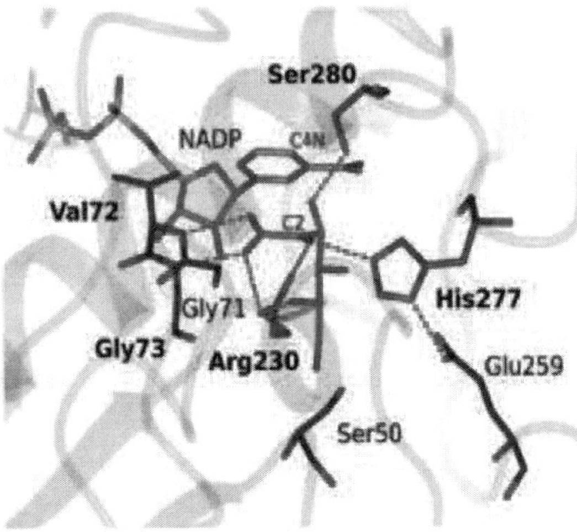

Figure 1. The catalytic site of ternary complexes of *Sm*GhrB with 2-keto-D-gluconate and NADP [Kutner et al., 2018]

S07-5
Approach towards stabilization of misfolded lossoffunction variants of human prolidase

P. Wilk[1], E. Wątor[1], M. Rutkiewicz[2], M. S. Weiss[3]
[1]Jagiellonian University, Małopolska Centre of Biotechnology, Kraków, Poland
[2]Max-Delbrück-Centrum für Molekulare Medizin, Berlin, Germany
[3]Helmholtz-Zentrum Berlin für Materialien und Energie GmbH, Berlin, Germany

Prolidase is the only metalloenzyme in humans capable of hydrolyzing dipeptides with proline or hydroxyproline on its Cterminus. The catalytic activity of the enzyme is crucial for proper recycling of proline in the system and patients affected by the syndrome referred to as Prolidase Deficiency (PD) suffer due to a number of various clinical symptoms. PD is a very rare recessive hereditary disorder with an estimated occurrence less than 1 in 1 000 000 birth and no cure to PD is available so far. The cause of PD is mutation in PEPD gene coding the enzyme. Almost 30 different mutations were reported to date and eight of them leading to either substitution of deletion of single amino acids were recently characterized structurally. Three of the mutants described displayed a significant degree of structural disorder leading to almost complete loss of catalytic activity.

In previous studies it was shown, that induction of Hsp70/90 expression in human fibroblasts leads to 2040% rescue of prolidase activity depending on the nature of mutation. To investigate this effect in more details we decided to structurally characterize three partly disordered prolidase variants expressed in the absence and in the presence of Hsp70/90 homologues.

In one of the variants (G278D) the side chain of the aspartate residue repulses and distorts the neighboring β-strand. If the variant was expressed in the presence of elevated levels of chaperones striking changes were observed and the resulting structure resembles the wildtype prolidase structure much more closely. Two more disordered variants were analyzed and just like in the previous studies the degree of stabilization was case dependent.

Our studies show that proteins expressed in the presence of chaperonins fold more tightly and better resemble the structure of wildtype prolidase.

We believe, that our findings will facilitate a new avenue leading ultimately to stabilization of misfolded lossoffunction proteins in affected patients.

S07-6
Aromatic amino acids aminotransferase from *Psychrobacter* sp. B6 - enzyme active site adaptability

A. Bujacz[1], J. Rum[1], M. Rutkiewicz[1]
[1]Łódź University of Technology, Institute of Molecular and Industrial Biotechnology, Łódź, Poland

The aromatic amino acids aminotransferase from *Psychrobacter* Sp. B6 (*Psy*ArAT) is the only one AroAT from psychrophilic organism with known crystal structure (PDB: 4RKC). It is a typical type-I of PLP-dependent enzyme, its fold is common for the most of aminotransferases, but it shows broader substrate specificity than the other known mesophilic and thermophilic ATs. It is an excellent example of cold adapted protein, which can be used in chemical and pharmaceutical industry to produce, among others, an enantiomerically pure products with a high efficiency.

Crystal structures of *Psy*ArAT in complexes with four substrate hydroxy-analogs of: tryptophan (2.6 Å), phenylalanine (2.3 Å), tyrosine (2.5 Å) and aspartic acid (1.6 Å), show in enzyme active center conformational changes, resulting from the presence of different sizes of ligands. Hydroxy-analogs of amino acids in which α-amino group was substituted by hydroxyl group act as competitive inhibitors - bind in active center but do not undergo transamination reaction, which makes them outstanding tool for enzyme catalytic center examination.

Structural studies of this enzyme, co-crystallized with substrates hydroxy-analogs provide a precious knowledge about interactions with ligand in an active center and conformational changes of the enzyme present in protein-ligand complexes.

This research was supported by grant 2016/21/B/ST5/00555 (A.B.) from the National Science Centre, Poland

References

Bujacz A., Rutkiewicz-Krotewicz M., Nowakowska-Sapota K., Turkiewicz M., *Acta Crystallographica D*, 2015, 71, 532-645.

Organic molecules and coordination compounds I

S08-1

Ab initio structure prediction of metal-organic frameworks

M. Arhangelskis[1], J. P. Darby[2], A. D. Katsenis[3], J. M. Marrett[3], T. Friščić[3], A. J. Morris[4]
[1]University of Warsaw, Faculty of Chemistry, Warsaw, Poland
[2]University of Cambridge, Cavendish Laboratory, Cambridge, United Kingdom
[3]McGill University, Department of Chemistry, Montreal, Canada
[4]University of Birmingham, School of Metallurgy and Materials, Birmingham, United Kingdom

Metal-organic frameworks (MOFs) are highly versatile crystalline microporous materials with applications in gas storage and separation, catalysis and light harvesting. The diversity in applications of these materials stems from the modular nature of MOFs, where the multitude of node and linker combinations lead to a vast structural space. But how can we navigate this space in a quest for materials with desirable properties?

Currently design of new MOFs is dominated by experimental screening, where multiple candidate structures are synthesized and tested for the property of interest. Naturally, such an approach is laborious, expensive and prone to the generation of excessive chemical waste. It would be highly desirable to develop a computational method for screening the potential MOF structures and envisage structures with the most promising properties for experimental synthesis.The progress in this avenue, however, is hampered by the lack of a reliable method for *ab initio* crystal structure prediction (CSP) of MOFs. This is in stark contrast with porous organic molecular materials, where CSP has been instrumental in advancing the theory-driven materials development.[1]

In this presentation we will report the first *ab initio* method for MOF structure prediction, and test it against a diverse set of MOFs, with differences in topology, metal coordination geometry and ligand binding sites. Our CSP approach is based on the *ab initio* random structure search (AIRSS)[2] method. The efficiency of structure generation is greatly enhanced with the introduction of Wyckoff alignment of molecules (WAM)[3] which uses the point group symmetry of molecular fragments to generate trial structures where such fragments can occupy high symmetry positions.

[1] Jones, J. T. A. et al. *Nature* 2011, *474*, 367.
[2] Pickard, C. J.; Needs, R. J. *J. Phys. Condens. Matter* 2011, *23*, 053201.
[3] Darby, J. P.; Arhangelskis, M. et al. *ChemRxiv* 2019 DOI:10.26434/chemrxiv.8204159

S08-2

Calcium Acetate Hydrates: Simple Salts with Surprisingly Complex Crystal Structures

S. Bette[1], M. X. Müller[2], G. Eggert[3], T. Schleid[2], R. E. Dinnebier[1]
[1]Max-Planck-Institute for Solid State Research, Scientific Facility X-ray diffraction, Stuttgart, Germany
[2]University of Stuttgart, Institute of Inorganic Chemistry, Stuttgart, Germany
[3]State Academy of Art and Design, Stuttgart, Germany

Introduction

Calcareous museums objects are often affected by corrosion phenomena induced by acetic acid vapor leading to the formation of efflorescence salts, most frequently calcium acetates. In the binary system $Ca(CH_3COO)_2$-H_2O the existence of three polytypes of $Ca(CH_3COO)_2 \cdot H_2O$, $Ca(CH_3COO)_2$ and a single form of $Ca(CH_3COO)_2 \cdot 0.5H_2O$ are known, but only the crystal structures of calcium acetate monohydrate polymorphs have been solved yet.

Objectives

In order to complete the structure knowledge of the binary system $Ca(CH_3COO)_2$-H_2O and to obtain reliable reference data for phase identification the crystal structure of $Ca(CH_3COO)_2 \cdot 0.5H_2O$ and the thermal decomposition of $Ca(CH_3COO)_2 \cdot H_2O$ leading to the anhydrous calcium acetate was investigated.

Results

$Ca(CH_3COO)_2 \cdot 0.5H_2O$ was obtained by a model experiment simulating the crystallization of efflorescence salts on porous objects (Fig. 1, a-c). Indexing of the powder pattern led to a surprisingly large unit cell with 11794.5(3) Å³. By *ab initio* structure determination a triple helix motif was revealed (Fig. d-g). Hence the structure of the hemihydrate (α-$Ca(CH_3COO)_2 \cdot 0.5H_2O$) exhibits striking similarities to the collagen proteins. A temperature dependent *in situ* X-ray powder diffraction study on the thermal decomposition of $Ca(CH_3COO)_2 \cdot H_2O$ revealed the existence of further calcium acetate sub-hydrates (β-$Ca(CH_3COO)_2 \cdot 0.5H_2O$ and $Ca(CH_3COO)2 \cdot 0.25H_2O$) and the existence of three anhydrous calcium acetate polymorphs (γ-, β- and α- $Ca(CH_3COO)_2$) with the latter one showing also a high- and a low-temperature form.

Conclusions

The simple binary system $Ca(CH_3COO)_2$-H_2O exhibits a surprisingly large variety of solid phases with complex crystal structures.

Fig. 1. $Ca(CH_3COO)_2 \cdot 0.5H_2O$: a) crystallization; b, c) SEM-images; d, e) triple helical motif; f, g) packing diagrams of right-handed (R) and left-handed (L) triple helices.

Figure 1

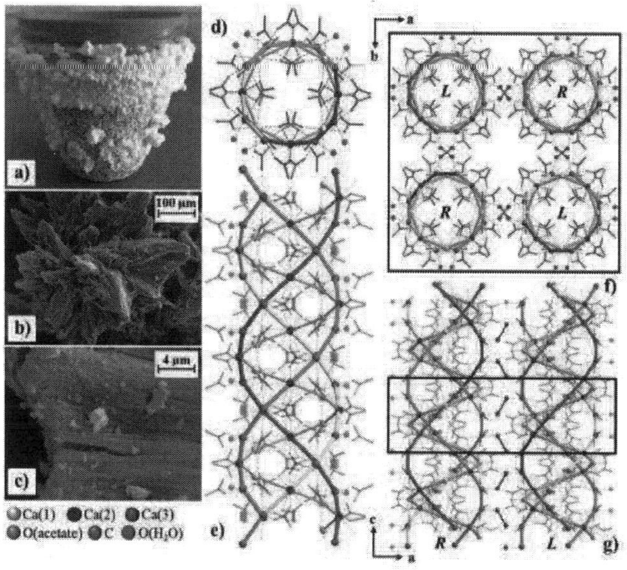

S08-3

Giant Supramolecules Meet Synchrotron Radiation: Experience with DESY P11 and P24 Beamlines

A. Virovets[1,2], E. Peresypkina[1,2], M. Scheer[1]
[1]University of Regensburg, Institute of Inorganic Chemistry, Regensburg, Germany
[2]Novosibirsk State University, Novosibirsk, Russian Federation

During last decades we have been using pentaphosphaferrocenes, $[Cp^RFe(\eta^5-P_5)]$ ($Cp^R = \eta^5-C_5R_5$, R = Me, CH_2Ph, PhC_4H_9, *etc*) as building blocks for the rational design of giant supramolecules, up to 4.6 nm in size [1-3]. The supramolecules consisting of hundreds of atoms frequently demonstrate weak scattering power due to the severe crystallographic disorder. In many cases, the use the high-flux synchrotron sources becomes the only remedy.

To obtain quality data at *d*min > 1 Å we perform experiments at DESY PETRA III P11 and P24 beamlines. For P11 beamline, the most critical is to choose radiation energy as a compromise between *d*min and quantum efficiency (QE) of a PILATUS 6M detector. The higher energies improves the resolution by the cost of significantly lower QE.

Optics at P24 beamline allows using hard X-ray radiation with E up to 44 keV. It helps reducing the absorption in Ag and Ta containing crystals and presumably radiolysis. The helium open-flow cryostat provides temperature down to 10 K. Sample preparation room is equipped with the vacuum-argon line within the fume hood to work with air-sensitive organometallic samples.

Optimization of the experimental strategy allowed us obtaining high-quality diffraction data even from weakly scattering crystals (Fig. 1) and investigate such subtle structural effects as superstructural ordering.

This work was supported by the SELFPHOS grant ERC AdG-339072.

Fig. 1a) Giant cationic supramolecule $[(Cp''Fe(\eta^5-P_5)_{12}(CuNCCH_3)_8]^{8+}$ with an external diameter of 2.5 nm. Hydrogen atoms are omitted for clarity; b) the diffraction pattern from the single crystal at P11 beamline (E = 18 keV, 0.1° scan)

[1] E. Peresypkina, C. Heindl, A. Virovets, M. Scheer (2016) *Structure and Bonding* **174**, 321
[2] H. Brake, E. Peresypkina, C. Heindl, et al (2019) *Chem. Sci.* **10**, 2940
[3] E. Peresypkina, C. Heindl, E. Mädl, et al (2018) *Chem.-A Eur. J.* **24**, 2503

Figure 1

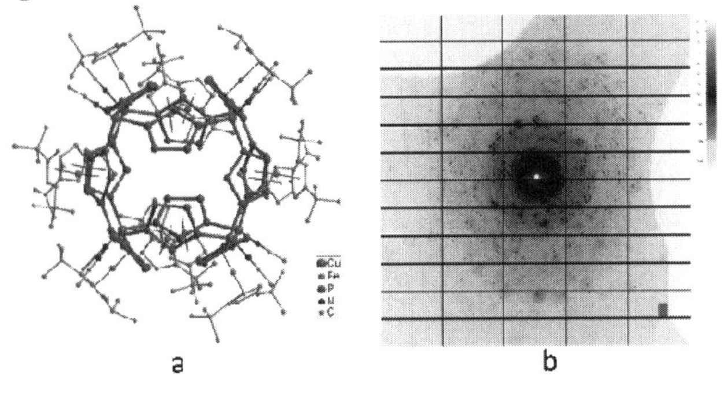

a b

S08-4

Crystal Structure Engineering with FlexCryst by Visualization of the Intermolecular Interactions

D. W. M. Hofmann[1], L. Kuleshova[2]
[1]CRS4, Modeling & Simulation , Pula, Italy
[2]FlexCryst, Uttenreuth, Germany

Already in 1971 the ideas of crystal engineering have been introduced. In the beginning the crystal structures have been discussed mainly in terms of atom pair distances and hydrogen bonds. Later Desiraju extended the considerations to more general schemes, which he named synthons. In the sense of machine learning descriptors have been introduced to describe crystal structures and by counting the synthons one obtains a frequency vector. However, the different descriptors didn't have a weight and their importance have been only estimatetd according statistics. To get more accurate vlaues for the importance we did introduce data mining (1,2) in crystallography. This allows to give exact weights for the different contributions.

The presented force field is trained on 100.000 structures and contains force field parameters for all atoms (in contrast to the universal force field). The visualization of the pair interactions give us a better insight in the forces acting within crystals. During our work the most important application is the finding of faulty structures. Presently the obtained force field is accurate enough to recognize the small inaccuracies, which are intrinsic for crystal structures solved from powder diagrams. Further applications are the targeted substitution of groups or atoms to achieve a more dense packing, important for explosives, or to lower the interaction energy to achieve higher solubility.

1) DWM Hofmann and T Lengauer. "Prediction of crystal structures of organic molecules." 474.1-3 (1999): 13-23.
2) DWM Hofmann, LN Kuleshova, eds. Data mining in crystallography. Vol. 134. Springer, 2009.

Figure 1: Crystal structure of a cobalt complex determined from single crystal structure. The stability of the dimer is counterbalanced by attractive and repulsive interactions.
Figure 2: Crystal structure of a cobalt complex determined from the powder diagram. The misplaced hydrogen becomes obvious by the strong repulsive interaction (15.4 KJ/mole).

Figure 1

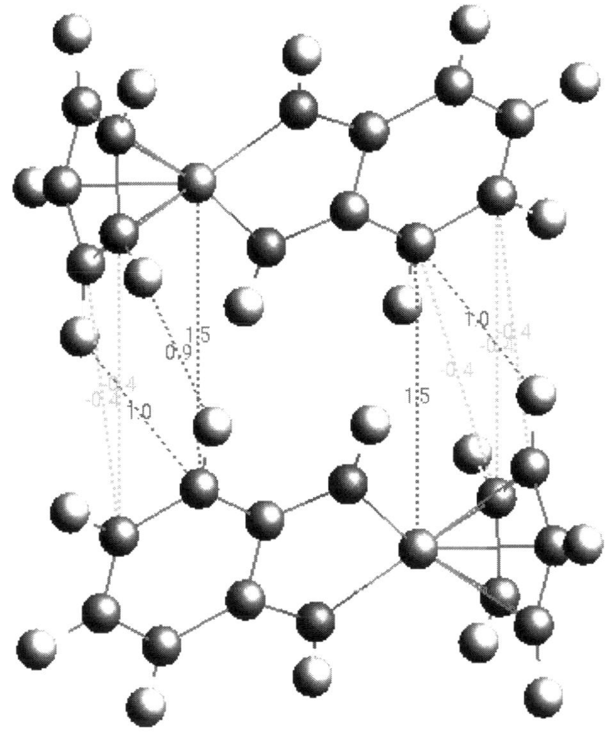

Figure 2

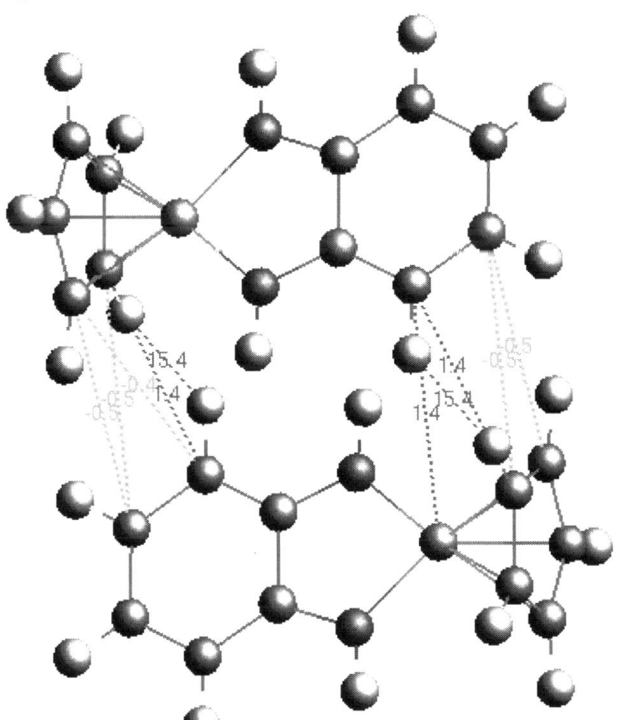

S08-5

Synchrotron-based structural studies of coordination compounds with interesting magnetic properties

T. M. Muzioł[1], R. Podgajny[2], N. Tereba[1], G. Wrzeszcz[1]
[1]Nicolaus Copernicus University, Faculty of Chemistry, Toruń, Poland
[2]Jagiellonian University, Faculty of Chemistry, Kraków, Poland

A synchrotron radiation is widely used in structural studies not only of macromolecules but also the most challenging small molecules possessing interesting properties and features: magnetic, fluorescent, conductive, chirality and structural conversions. In magnetic compounds paramagnetic ions are connected by properly selected bridging ligands enabling efficient magnetic coupling (cyanides) and abundance of the observed coordination modes (oxalates) [1]. The topology of these complexes can be controlled and modified by proper selection of auxiliary ligands.

We aimed to use synchrotron radiation for structural studies of the most challenging coordination compounds. The diffraction data sets for the most challenging compounds showing interesting magnetic properties were collected at MX beamlines of BESSY II (HZB, Berlin) synchrotron. The diffraction experiments were performed for unstable compounds degrading quickly in time due to external conditions (e.g. solvent loss, structural conversion), showing weak diffraction and/or possessing big cell parameters. Problems encountered during data collections and/or refinement will be discussed as well as results of magnetic and theoretical studies will be presented. We discuss structure and magnetic properties of trimetallic (CoCuFe) 1D chain with antiferromagnetic coupling between two Cu ions reaching -250 cm^{-1} and topology of 15-centered Co_9W_6 cores based on octacyanotungstate controlled by N,O-ligands. The results of XAS studies performed at PEEM/XAS beamline of National Synchrotron Radiation Centre SOLARIS (Kraków) proved structure conversion occurring between oxalate complexes.

The obtained results prove that experiments performed with synchrotron radiation provide valuable structural information enabling modelling of magnetic properties, important for design of new magnetic materials.

References

[1] T.M. Muzioł, N. Tereba, R. Podgajny, D. Kędziera, G. Wrzeszcz, *Dalton Trans.*, **2019**, 48, 11536–11546.

S08-6

Photocrystallographic studies of a series of novel nickel(II) nitro complexes in the crystal state

P. Borowski[1], S. Kutniewska[1], R. Kamiński[1], K. Jarzembska[1]
[1]University of Warsaw, Department of Chemistry, Warsaw, Poland

The importance of solid-state materials of real-life applications is constantly increasing. In this view, conscious design of new functional photoswitchable chemical systems cannot be overestimated. Molecular switches triggered by light may find applications in optoelectronics, colour-changing materials, high-capacity storage devices, etc. Transition-metal complexes, in which metal centre is coordinated by molecular fragments that can exist in multiple isomeric forms, are among potential functional materials of this kind.

Hence, the current study was devoted to profound investigations of newly designed and synthesised photoactive Ni(II) complexes, which contain the NO_2 group as the ambidentate ligand. This moiety may undergo an isomerisation reaction upon light irradiation and/or temperature changes. The primary aim of the project was to examine the photoreactivity of the obtained crystal systems and understand the nitro-nitrito linkage isomerisation mechanism. For that purpose multi-temperature (photo)crystallographic and spectroscopic experiments were conducted and supplemented by computational analyses.

All studied compounds consist of the Ni(II) nickel centre chelated by a (N,N,O)-donor ligand and the nitro moiety. They crystallise either in P(-1) or $P2_1/c$ space groups. The *nitro* linkage isomer switches (20-33% conversion) to the *nitrito* form upon specific LED irradiation and temperature (530/590nm at 140/160K) indicated by the precedent IR spectroscopy measurements. It is also worth noting that the process can be triggered by the temperature itself, however, with significantly lower efficiency. The metastable species exist in the 140 – 190 K range, while above 190 K the complex reverts back to its ground-state form.

PRELUDIUM (2017/25/N/ST4/02440), WCCS (grant No. 285) and European Regional Development Fund (POIG.02.01.00-14-122/09) programs are acknowledged.

[1] Kamiński et al., *J. Appl. Cryst.* 2016, 49, 1383
[2] Hatcher et al., *Acc. Chem. Res.* 2019, 52, 4, 1079

Figure 1

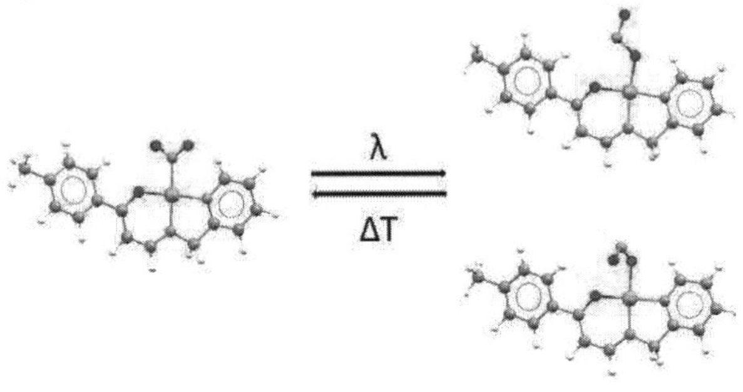

Bio-Crystallography III: Instrumentation & hybrid methods

S09-1

Crystallographic research opportunities at Polish synchrotron - NSRC SOLARIS

M. Stankiewicz[1], M. Kozak[2,1], T. Kołodziej[1], A. Wawrzyniak[1], M. Rawski[3], P. Indyka[3], K. Wrobel[3], S. Glatt[3]
[1]Jagiellonian University, National Synchrotron Radiation Centre SOLARIS, Kraków, Poland
[2]Adam Mickiewicz University, Department of Macromolecular Physics, Poznań, Poland
[3]Jagiellonian University, Małopolska Centre of Biotechnology, Kraków, Poland

SOLARIS National Synchrotron Radiation Centre at the Jagiellonian University (Kraków, Poland), is the first synchrotron, which has been built in Poland, as well as in Eastern Europe. The heart of this facility is a low energy storage ring (E=1.5 GeV, circumference 96 m) constructed in a modern, integrated Double-Bend Achromat (DBA) technology. Currently, several beamlines dedicated to XAS, UARPES and XPS spectroscopies are operational for users or are in advanced construction stage at Solaris storage ring.

In response to the expectation of the crystallographic community, the construction of SOLCRYS beamline, based on a superconducting wiggler and operating in the energy range up to 24 keV, has also begun. This beamline will have two endstations (one dedicated to macromolecular crystallography and second to SAXS and general XRD experiments) and will be located in the extended part of the existing experimental hall. It is being constructed in close collaboration with Joint Institute for Nuclear Research (JINR) in Dubna, Russia.

Since December 2019, NCSR Solaris also offers application-based access to a Titan Krios 3Gi cryo electron microscope (Cryo-EM). Another Cryo-EM for screening purposes and Micro-Electron Diffraction (MicroED) experiments will be installed later this year. The presentation will summarise current and future research opportunities in the field of crystallography at NSRC Solaris.

S09-2

Structural Biology at the refurbished European Synchrotron

C. MUELLER-DIECKMANN[1]
[1]European Synchrotron Radiation Facility (ESRF), Structural Biology, Grenoble, France

The European Synchrotron (ESRF) has undergone a complete refurbishment of its machine and storage ring. This new setup, called ESRF-EBS, changes the source characteristics and converts the ESRF to a fourth generation synchrotron site. The presentation given will give an overview of the macromolecular crystallography (MX) beamlines available to ESRF"s international users" community and how ESRF-EBS will help to better tackle challenging experiments (e.g. inter- and intra- crystal variability with respect to diffraction quality, multi crystal data collection strategies etc.). Additionally, facilities complementary to MX as well as a new beamline dedicated to synchrotron serial crystallography, currently under construction, will be covered.

S09-3

Biocrystallography at Beamline P11

J. Hakanpää[1], E. Crosas[1], S. Saouane[1], J. Meyer[1], J. Urbschat[1], B. Reime[1], A. Meents[2], A. Burkhardt[1]
[1]Deutsches Elektronen Synchrotron (DESY), Photon Sciences, Hamburg, Germany
[2]Center for Free Electron Laser Science, DESY, Hamburg, Germany

Beamline P11 at PETRA III in Hamburg is a versatile instrument for macromolecular crystallography (1). The photon energy can be adjusted between 5.5 - 28 keV with the possibility of using a CdTe-detector for higher energies (> 22 keV). Beam sizes between 200 x 200 μm and 4 x 9 μm are available, with a maximum photon flux of 1e13 ph/s at 12 keV.

P11 is optimized for high-throughput crystallography. Typical data collection times are <2 min. In spring 2020, the Pilatus 6M detector will be replaced by an Eiger2 16M. The automatic sample changer at P11 is based on unipuck format. 23 pucks (368 samples) can be loaded into the robot dewar simultaneously. The mounting cycle is 20 s. Remote access is prepared for the users in mid-2020.

Serial data collections are implemented either as fast 2D scans in the graphical user interface or as series of rotation wedges in a separate scan tool. In the near future, the installation of the Roadrunner goniometer (2) is planned in order to further speed-up the serial scans and allow for pharmaceutical applications *e.g.* fast screening of ligand libraries. P11 is compatible with the micropatterned chips (3), which can hold thousands of crystals and have pore sizes between 2 and 50 μm.

Owing to the flexible setup of the beamline, various non-standard sample delivery systems and devices can be implemented at P11 *e.g* the tape-drive set-up for serial crystallography, in which a suspension of crystals (typically sub 15 μm in size) are written on a polyimide tape and then delivered to the beam. This system enables time-resolved experiments by the mix-and diffuse method (4).

The call for regular proposals is open twice a year via the DESY DOOR system. Travel reimbursement can be applied for by researchers from German Universities, EU and EU-associated countries.

[1] Burkhardt et al. (2016) Eur. Phys. J. Plus ***131***:56.
[2] Meents et al. (2017) Nat. Commun ***8***:1281.
[3] Roedig et al. (2015) Sci. Rep. ***5***: 10451.
[4] Beyerlein et al. (2017) IUCrJ ***4***:769.

S09-4

T-REXX: PETRA-III's NEW ENDSTATION FOR SERIAL TIME-RESOLVED CRYSTALLOGRAPHY

M. Agthe[1], D. von Stetten[1], G. Bourenkov[1], M. Polikarpov[1], S. Horrell[2], B. Yorke[3], G. S. Beddard[4], M. Nikolova[1], I. Karpics[1], T. Gehrmann[1], J. Meyer[1], U. Ristau[1], S. Fiedler[1], D. C. F. Monteiro[5], M. Trebbin[6], P. Mehrabi[7], E. C. Schulz[7], F. Tellkamp[7], R. J. D. Miller[7], N. Huse[8], A. R. Pearson[8], T. R. Schneider[1]
[1]European Molecular Biology Laboratory (EMBL) , Hamburg, Germany
[2]Diamond Light Source, Didcot, United Kingdom
[3]University of Leeds, Leeds, United Kingdom
[4]University of Edinburgh, Edinburgh, United Kingdom
[5]Hauptman-Woodward Medical Research Institute, Buffalo, United States
[6]University at Buffalo, Buffalo, United States
[7]Max Planck Institute for the Structure and Dynamics of Matter, Hamburg, Germany
[8]University of Hamburg, Hamburg, Germany

EMBL operates two beamlines (P13 and P14) for macromolecular crystallography at the PETRA III synchrotron in Hamburg. Recently, we have built a new endstation (P14.2, T-REXX) for serial time-resolved crystallography as an extension to P14 that has been in operation since October 2018. For this endstation, the P14 X-ray beam is refocused via a series of X-ray refractive lenses mounted in a transfocator to provide a 15 x 10 μm² beam with a flux of ca. $2 \cdot 10^{12}$ photons/s at an energy of 12.7 keV.

T-REXX is modularly designed for serial crystallography with support for various microcrystal delivery systems and is therefore not equipped with a goniometer. Instead, we use a basic beam conditioning unit, also called beam shaping device (BSD) that provides an on-axis viewing system, beam shaping apertures, a fully motorized beamstop, and two scintillators to image X-ray and laser beams. Our pre-installed laser system emits UV light at a wavelength of 355 nm and can be used to initiate reactions in protein crystals. The nearly co-axial coupling of the laser beam with the X-rays at the sample position simplifies alignment of the beams using the on-axis viewing system. First experiments using patterned silicon chips as well as microfluidic flow cells have yielded promising results. Currently, we aim at reaching sub-millisecond timescales in time-resolved experiments by using the Hadamard technique [1], either by gating the detector or alternatively by modulating the X-ray beam.

[1] Yorke BA et al., Nat Methods, 2014

S09-5

Neutron protein crystallography at the Heinz Maier-Leibnitz Zentrum (MLZ): New developments and recent application examples

T. E. Schrader[1], A. Ostermann[2], M. Monkenbusch[3], B. Laatsch[4], P. Jüttner[2], W. Petry[2], D. Richter[1,3]
[1]Forschungszentrum Jülich GmbH, Jülich Centre for Neutron Science, Garching, Germany
[2]Technical University Munich, Heinz Maier-Leibnitz Zentrum (MLZ), Garching, Germany
[3]Research Center Jülich GmbH, Institute for Complex Systems ICS, Jülich, Germany
[4]Research Center Jülich GmbH, Engineering and Technology (ZEA-1), Jülich, Germany

The neutron single crystal diffractometer BIODIFF at the research reactor Heinz Maier-Leibnitz (FRM II) is especially designed to collect data from crystals with large unit cells. The main field of application is the structural analysis of proteins, especially the determination of hydrogen atom positions. BIODIFF is a joint project of the Jülich Centre for Neutron Science (JCNS) and the FRM II. BIODIFF is designed as a monochromatic instrument with a narrow wavelength spread of less than 3 %. To cover a large solid angle the main detector of BIODIFF consists of a neutron imaging plate in a cylindrical geometry with online read-out capability.

BIODFF is equipped with a standard Oxford Cryosystem "Cryostream 700+" which allows measurements at 100 K. A new kappa goniometer head was added recently. This allows an automated tilting of the crystal in order to increase the completeness of the data set when recording another set of frames in the tilted geometry. Typical scientific questions addressed are the determination of protonation states of amino acid side chains in proteins and the characterization of the hydrogen bonding networks between the protein active centre and an inhibitor or substrate.

Picking out some recent highlights from measurements at BIODIFF it will be shown how the method of neutron protein crystallography could be used to answer mechanistic questions in enzymatic processes or help to improve inhibitor fragment screening. New developments at the instrument will also be presented: A new collimation for the primary beam should lead to a reduction in background. It should also make it easier to align the neutron beam with the centre of the neutron imaging plate detector.

S09-6

Construction of highly ordered materials composed of protein containers and plasmonic nanoparticles

M. Lach[1], T. Beck[1]
[1]University of Hamburg, Institute of Physical Chemistry, Hamburg, Germany

We have introduced a method for the construction of multifunctional biohybrid material, using protein containers, engineered with opposite surface charge[1], as atomically precise ligand shells for the assembly of inorganic nanoparticles into highly ordered superlattices[2]. The engineered protein containers form crystalline assemblies, which were characterized with single crystal X-ray crystallography to high resolution. In previous works the nanoparticles were synthesized in situ inside of the protein container cavity. Here, the nanoparticle loaded protein container building block is obtained by a dis- and reassembly approach[3]: the protein container is disassembled by acidic or chaotropic conditions, and pre-synthesized plasmonic nanoparticle are encapsulated during the reassembly of the protein container. The crystallization of oppositely charged protein containers with nanoparticle cargo yields highly ordered nanoparticle superlattices as free-standing crystals, with up to a few hundred micrometers in size. The artificially assembled materials will be able to manipulate light at the nanoscale, which is important in metamaterials and for applications such as sensing, plasmon optics or light harvesting.

References:

[1] T. Beck, S. Tetter, M. Künzle, D. Hilvert, *Angew. Chemie Int. Ed.* 2015, *54*, 937–940.
[2] M. Künzle, T. Eckert, T. Beck, *J. Am. Chem. Soc.* 2016, *138*, 12731–12734.
[3] M. Künzle, J. Mangler, M. Lach, T. Beck, *Nanoscale* 2018, *10*, 22917-22926.

Figure 1

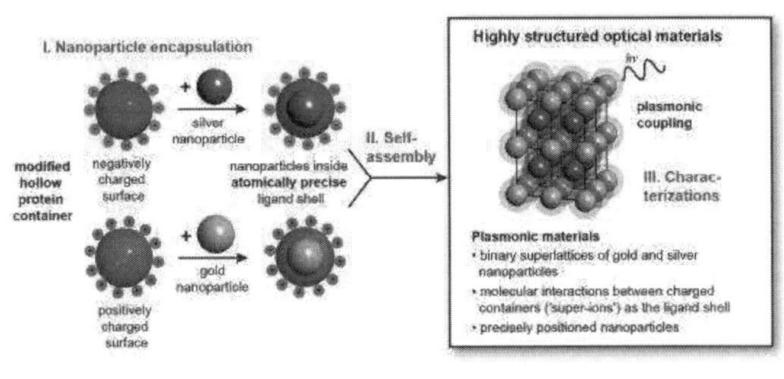

Structure-property-relationships II

S10-1
Structure-property relationships in a reinforced calcium phosphate cement based on metastable α′-tricalcium phosphate

A. Goncharenko[1], Z. Zyman[1], M. Epple[2], O. Onyshchenko[1], O. Prymak[2], K. Loza[2]
[1]V.N. Karazin Kharkiv National University, Physics of Solids Department, Kharkiv, Ukraine
[2]University of Duisburg-Essen, Inorganic Chemistry and Centre for Nanointegration Duisburg-Essen (CeNIDE), Essen, Germany

Introduction
Here we present new features of the factors determining the cement setting process and structural and mechanical characteristics of a cement based on metastable α-tricalcium phosphate (α′–TCP) and reinforced with hydroxyapatite (HA) whiskers.

Materials and methods
α′–TCP was prepared by a modified fast nitrate synthesis [1]. The formed precipitate was filtered and subjected to consolidation in an aqueous medium (3 L, 5 °C, pH 9, 3 h), then filtered and lyophilized (-23 °C, 24 h). The obtained amorphous powder was heated in air (5 K/min), crystallized at 700 °C for 1 h and hardened by fast cooling. HA whiskers were synthesized by a hydrothermal method [2]. They were blended with a 2.5 wt% Na2HPO4 solution. The solutions were used as the liquid components of the α′–TCP powder in cements with a powder to liquid (P/L) ratio of 0.66 [3]. The products were analyzed by XRD, TG/DTA, IR, SEM, EDX, and mechanical testing. The setting kinetics of the cement were evaluated by the Vicat needle method.

Result and discussion
All the cements set within 15 min which is a realistic time for a surgical operation. The nanocrystalline state of the set cement is a favorable feature because it promotes the cement dissolution and the transformation into the bone tissue *in vivo*. The compressive strength also increased with higher whisker content from 3.5 MPa (no whiskers) up to about 7.5 MPa (4 wt% whiskers). Besides the mechanical reinforcement, the presence of the whiskers stimulated the setting processes. Since the compressive strength of cancellous bone at a typical porosity of 65% is in the range of 2–80 MPa, the cement seems to be usable in clinical practice.

Conclussion
A prospective apatite calcium phosphate cement based on metastable α′–TCP and reinforced by HA whiskers is presented.

References
[1] Zyman ZZ et al., J Mater Sci Mater Med, 21:123–130 2010
[2] Zyman ZZ et al., Biomaterialien, 7:252 2006
[3] Goncharenko A et al., ESB 2019, Abstracts: 893–894 2019

S10-2
Cu/Zn disorder vs. solar cell efficiency: the $Cu_2ZnSn(S_xSe_{1-x})_4$ monograin case

G. Gurieva[1], A. Franz[1], K. Muska[2], K. Ernits[2], S. Schorr[1,3]
[1]Helmholtz-Zentrum Berlin für Materialien und Energie GmbH, Structure and Dynamics of Energy Materials, Berlin, Germany
[2]crystalsol OÜ, Tallin, Estonia
[3]Freie Universität Berlin, Institute of Geological Sciences, Berlin, Germany

Kesterite-type based thin films solar cell technologies are mainly based on polycrystalline absorber layers. A promising low cost alternative technology uses kesterite monograins (single crystals of 50-100 µm size) which are fixed in a polymer matrix to form a flexible solar cell [1].

It is agreed in literature that large band tailing observed in Cu-based kesterite-type semiconductors causes voltage losses limiting the efficiency of kesterite-based devices [2]. The Cu/Zn disorder (CuZn and ZnCu anti-sites in Cu-Zn planes at z=¼ and ¾), which is always present in these compounds [3], is discussed as a possible reason for band tailing.

The experimental determination of the order parameter Q which is a quantitative measure of the degree of Cu/Zn disorder [4] requires a differentiation between the isoelectronic cations Cu+ and Zn2+. An in depth analysis of neutron diffraction data provides information on the cation distribution in the crystal structure allowing the determination of type and concentration of intrinsic point defects including a distinction between Cu and Zn [3]. On the other hand neutron diffraction requires large sample volumes, thus kesterite monograins offer the unique possibility to correlate structural disorder in kesterite-type absorbers with solar cell performance parameters.

We will present a detailed structural investigation of CZTSSe monograins based on neutron powder diffraction experiments. The order parameter Q representing the Cu/Zn disorder as well as the occurring intrinsic point defects have been determined. We will present the influence of the purity of the starting material (copper) as well as small changes in the chemical composition on the Cu/Zn disorder resulting in different power conversion efficiencies of the respective devices.

[1]www.crystalsol.com
[2] Rey et al., Sol. En. Mat. Sol. Cells 172 (2018) 149
[3] Gurieva et al., J. Appl. Phys. 123 (2018) 161519
[4] Toebbens et al. Phys. Stat. Sol. B 253 (2016) 1890

S10-3
Structural Versatility in Methylhydrazinium Lead Halides; New Perovskites with Exceptional Optical and Dielectric Properties

M. Mączka[1], A. Gągor[1], M. Ptak[1], D. Stefańska[1], A. Sieradzki[2]
[1]Institute of Low Temperature and Structure Research, Polish Academy of Sciences, Wrocław, Poland
[2]Wrocław University of Science and Technology, Faculty of Physics, Wrocław, Poland

Organic-inorganic perovskites of general composition $[AmH]PbX_3$ ($X=Cl^-,Br^-,I^-$) attracts unceasing attention due to the extraordinary performances in solar cell applications, intriguing order-disorder phase transitions and useful electronic properties. Hybrid perovskites are expected to form for Goldschmidt's Tolerance Factor (TF) between 0.8 and 1. Giving that PbX_3^- framework may form only with the smallest organic cations. Despite the large research in this field, 3D structures have been reported so far only for MA^+ and FA^+. Other small amines crystallize in non-perovskite or layered perovskite-like structures.

Herein, we report a single-crystal x-ray diffraction, dielectric and optical properties of two new lead halides, based on protonated methylhydrazine (MHy^+) which break out of the rules. First one, $MHyPbBr_3$ features TF higher than 1 (1.03) and crystalize in 3D structure, having non-centrosymmetric monoclinic $P2_1$ space group, fig.1. It preserves polar properties up to 418K where the phase transition to archetypal cubic *Pm-3m* phase takes place. The polar phase is greatly distorted, displays strong second-harmonic generation, switchable dielectric behavior, thermochromism and two-photon energy up-conversion under 800 nm excitation [1]. The second compound, MHy_2PbI_4, crystallizes in 2D structure with single [100] lead-iodide octahedra slabs, fig.2. It shows an exceptionally short separation of adjacent lead iodide layers as well as unusually small interoctahedral tilting of PbI_6 octahedra, leading to bandgap of 2.20 eV. MHy_2PbI_4 undergoes three structural phase transitions: *Pmmn-Pmmn*(320K)-*Pccn*(298K)-*P-1*(262K) contributing to switchable dielectric properties. MHy^+ is the smallest organic cation in $[AmH]_2PbI_4$ [2]. Our results imply that the TF range for new functional 3D and 2D perovskites should be extended.

Fig.1. MHyPbBr3 (300K) Fig.2. MHy2PbI4 (100K)

[1] M.Mączka et al. JACS – submitted
[2] M.Maczka et al. Chemistry of Materials 2019 *31* (20), 8563-8575

Figure 1

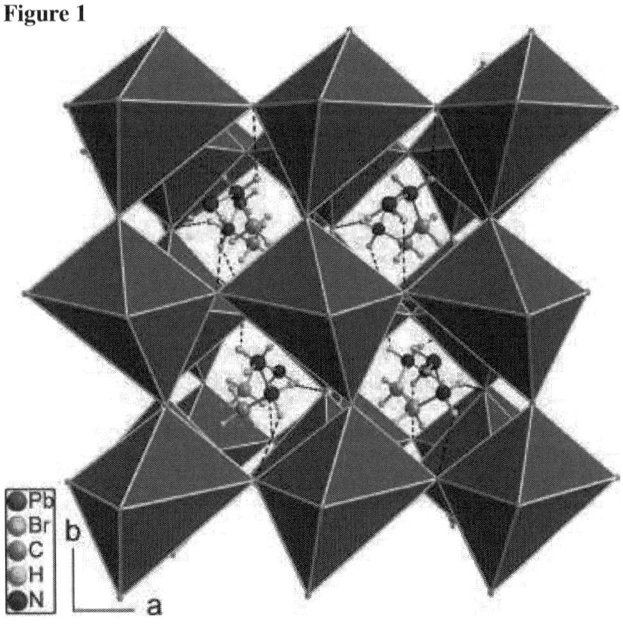

Figure 2

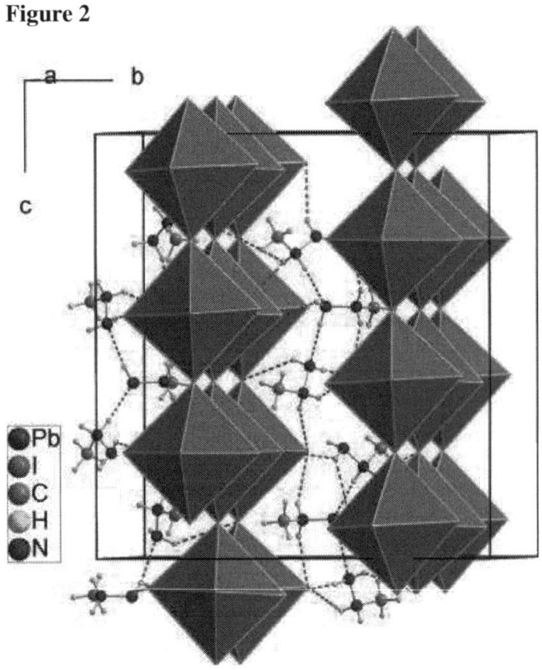

S10-4

Oxygen defect formation and ionic transport in $Nd_2CuO_{4\pm\delta}$ as derived from first principles calculations

B. Gędziorowski[1], K. Cichy[1], K. Świerczek[1]
[1]AGH University of Science and Technology, Energy and Fuels, Kraków, Poland

Introduction

Oxides with perovskite-related structures are in the interest of researchers and industry because they often show high mixed ionic electronic conductivity (MIEC) that allows for application in advanced electrochemical devices, e.g. for power generation or oxygen separation. Among the MIEC materials, oxides with general formula $A_2BO_{4\pm\delta}$ might be distinguished and considered as first member of the Ruddlesden-Popper $A_{n+1}B_nO_{3n+1}$ series. Their structure includes both occupied and unoccupied oxygen sites resulting in three structural variants with B cation in 4-, 5- and 6-fold coordination and symmetry changing from tetragonal to orthorhombic. The high degree of freedom for oxygen sites occupancy enables the possibility of effective oxygen ions transport by different mechanisms (vacancy, interstitial or interstitialcy) and via various diffusion paths.

Methods

Based on our previous studies [1], we employ first principles calculations (Materials Studio software with CASTEP code) to investigate oxygen transport properties in $Nd_2CuO_{4\pm\delta}$, the end member of $A_2BO_{4\pm\delta}$ series with I4/mmm symmetry and Cu ions in 4-fold coordination. We support theoretical predictions with experimental studies, including HT-XRD, thermogravimetry and oxygen permeation tests.

Results and conclusions

Our studies on Nd2CuO4±δ show that at moderate temperatures range interstitial oxygen defects are dominant with negative values of defect formation energy, whereas at elevated temperatures and low oxygen partial pressures oxygen vacancies are beginning to form within the CuO_2 layers. Analysis of all possible diffusion paths shows low activation energy in certain directions, for both interstitial and vacancy mechanism, which indicates good oxygen transport properties.

Acknowledgements
The project was funded by the National Science Centre, Poland on the basis of the decision number UMO-2015/19/B/ST8/00871.

References

[1] Z. Du, et. al, Phys. Chem. Chem. Phys. 20(33) (2018) 21685-21692

S10-5

Impact of crystal packing on the luminescence in crystalline forms of benzoyl-acetylide-gold(I) complexes

A. Makal[1], D. Plazuk[2], M. Głodek[2], S. Palwędzio[2]
[1]University of Warsaw, Department of Chemistry, Warsaw, Poland
[2]Łódź University of Technology, Łódź, Poland

Two-coordinated Au(I) complexes and the influence of Au···Au interactions on their luminescence and formation of polymorphs or solvates has drawn considerable attention in recent years. The ability to form various architectures of Au(I) complexes is attributed to the aurophilic interactions, which result in many types of complexes, differing in structure as well as in mechanical and luminescent properties of resulting materials.

Recently1 we reported on the influence of the crystal packing and aurophilic interactions of the gold(I) acetylide complexes of the type (ArCOC≡C)n AuPEt3 (n=1,2) on their luminescence.

In particular, (benzoylacetylide)gold(I) compound yielded four crystal forms with strikingly different luminescence. This was a rare opportunity where crystal structures and ensuing electronic properties of the same Au(I) complex, forming different sets of aurophilic interactions and placed in various crystalline environments, could be directly compared.

Crystal structures of all forms were determined with single crystal and powder X-ray diffraction. Structural analysis was complemented by periodic DFT calculations in order to relate structures to their observed spectroscopic properties. We also monitored the conversion pathway for these forms: an orange luminescent form of homoleptic complex upon drying underwent spontaneous transformation to a bright green fluorescent form and finally to the most stable, weakly blue emissive one. In addition, a rare example of a helical arrangement of Au···Au···Au chains was discovered in heteroleptic (benzoylacetylide)gold(I) complex.

The study showed that the shortest Au···Au distance observed in the crystal structure did not, by itself, determine the wavelengths or efficiency of solid-state luminescence. Kinked and helical Au···Au···Au chains yielded more efficient luminophores, irrespective of the direct Au···Au distances.

[1] M. Glodek, S. Pawledzio, A. Makal, D. Plazuk, Chem. Eur. J. 2019, 25, 13131-13145

S10-6

New insights in the catalytic activity of cobalt orthophosphate $Co_3(PO_4)_2$ from charge density analysis

D. Stalke[1]
[1]Georg-August Universität Göttingen , Department of Inorganic Chemistry, Göttingen, Germany

An extensive characterization of $Co_3(PO_4)_2$ was performed by topological analysis according to Bader"s Quantum Theory of Atoms in Molecules from the experimentally and theoretically determined electron density. This study sheds light on the reactivity of cobalt orthophosphate as a solid-state heterogeneous oxidative dehydration and dehydrogenation catalyst. The topological properties identified various faces of the bulk catalyst as possible reactive sites. The charge accumulations and depletions around the two independent five- and six-fold coordinated cobalt atoms, found in the topological analysis, are correlated to the orientation and population of the d orbitals. It is shown that the (011) face has structural features best for catalysis. Five-fold coordinated ions in close proximity to advantageously oriented vacant coordination sites and electron depletions suit the reactant"s oxygen-lone pairs most for chemisorption. This is confirmed both from the multipole refinement as well as from density functional theory calculations. Close-by basic phosphate ions are readily available for C–H activation.

Helena Keil, Matti Hellström, Claudia Stückl, Regine Herbst-Irmer, Jörg Behler, Dietmar Stalke "New insights in the catalytic activity of cobalt orthophosphate Co3(PO4)2 from charge density analysis" *Chem. Eur. J.* **2019**, *online*, DOI: 10.1002/chem.201902303.

Figure 1

Organic Molecules and coordination compounds II

S11-1
Hybrid inorganic-organic materials based on complexes of benzoilthioureas and group 11 or 12 metal salts

D. Rosiak[1], A. Okuniewski[1], J. Chojnacki[1]
[1]Gdansk University of Technology, Chemical Department, Gdańsk, Poland

Recently one of the fast developing branches of chemistry is research on coordination compounds, which form infinite structures. Further division of the structures based on the linker character is also necessary. With (only)organic linkers we have "coordination polymers", when inorganic atoms or groups link (bridge) coordination centres we can obtain "hybrid inorganic" chains, layers or 3D structures. Structures with mixed connectivity of various dimensions are generally known as "hybrid inorganic organic structures" I^nO^m, where n – dimension of inorganic substructure, m – dimension of organic substructure ($n + m \leq 3$). Infinite coordination compounds with inorganic bridges in polymeric structures may exhibit distorted electronic band structure of the skeleton which gives structures with desired catalytic, photovoltaic, luminescent or magnetic properties.

In our research we decided to explore bezoilthiourea complexes of group 11 and 12 metal salts since the metals are known from their catalytic activity and formation of inorganic-organic hybrid materials should influence their electronic structure by introduction of cooperative effects. We tried to find best conditions to synthesise infinite structures: chains, ladders, layers or 3D structures.

During the study we obtained and characterized structurally 30 ligands being substituted 1-bezoilthioures and 48 complexes with Cu, Hg, and Ag. Some of the structures have desired "infinite" topology forming chains, ladders or a quadruple chain.

Substituted benzoilthiourea molecules proved useful as stabilizing ligands for synthesis of hybrid inorganic chain or layer structures of halogens of group 11 and 12 metals. During the presentation examples of monomeric, dimeric, chain 1D or ladders (below) of the above complexes will be presented in more detail.

References

A. K. Cheetham, C. N. R. Rao, R. K. Feller, *Chem. Commun.*, **46** (2006) 4780

D. Rosiak, A.Okuniewski, J. Chojnacki, *Acta Cryst. C* **74** (2018), 1650-1655.

Figure 1

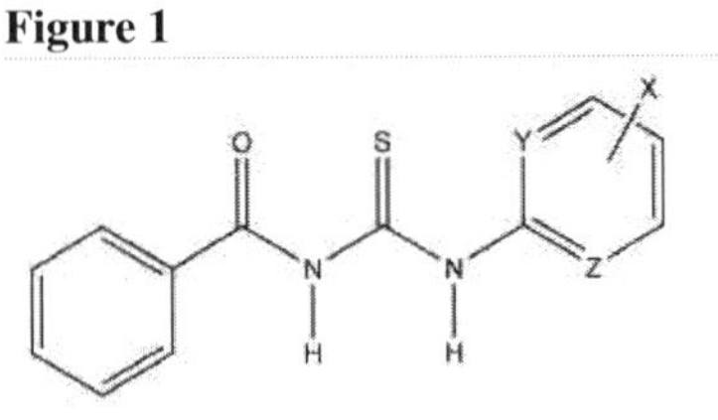

X – H, Me, Cl, Br, I, OH or OMe
Y, Z – N or CH

Figure 2

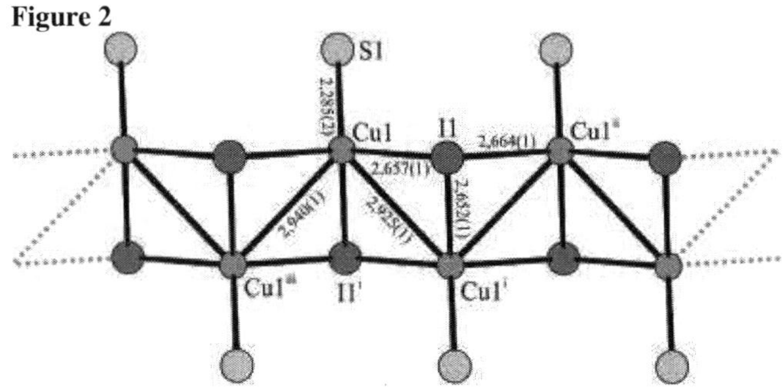

S11-2
Frustration and frustrated crystal structures of the *p-tert*-butylcalix[6]arene host.

M. Malińska[1]
[1]University of Warsaw, Department of Chemistry, Warsaw, Poland

Calix[n]arenes have applications in the areas of host-guest chemistry, ion and molecular recognition, enzyme mimics, catalysis, interaction with biomolecules, ion extraction, and selective ion transport. Crystallisation conditions (supersaturation, temperature, additives) were screened to obtain crystal structures with varying architecture such as layers, channels and spherical voids with different conformations of the host p-tert-butylcalix[6]arene using different guests (cyclohexane, anisole, cyclohexane, toluene, benzene, methyl acetate, dichloromethane, THF, pyridine), which were characterised by X-ray diffraction. The molecular recognition and self-assembly between host and guest (cyclohexane, pyridine, dichloromethane) led to kinetic trapping and crystallization of the intermediate complexes depending on the host-guest molecular volume ratio. The crystalline kinetic complex underwent slow spontaneous dissolution and subsequently recrystallizes as thermodynamic inclusion complexes. Potential of mean force calculation with the aid of umbrella sampling of the p-tert-butylcalix[6]arene conformational change in different solvents (guests) can explain diverse host-guest structural architecture.

Figure 1

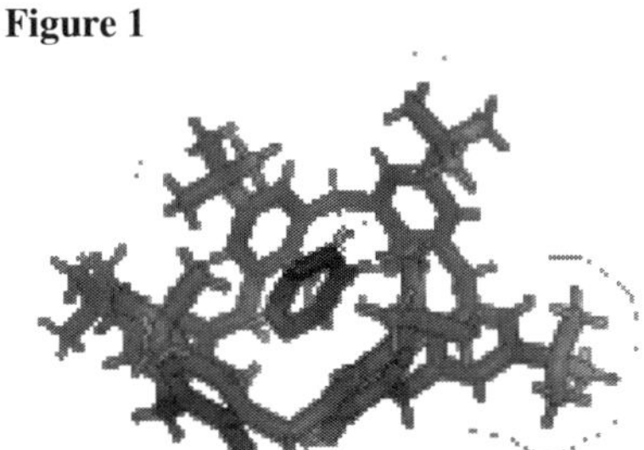

S11-3
Single Crystal X-ray and Neutron Diffraction Experiments in the Identification of Non-Innocent Methylene Linker in Bridged Lewis Pair Initiators

M. Weger[1], R. Grötsch[1], M. Knaus[1], M. Giuman[1], D. Mayer[1], P. Altmann[1], E. Mossou[2], B. Dittrich[3], A. Pöthig[1], B. Rieger[1]
[1]Technische Universität München, Munich, Germany
[2]ILL, Grenoble, Germany
[3]Heinrich Heine University, Düsseldorf, Germany

Deprotonation usually is an unwanted side reaction in the Lewis pair polymerization of Michael acceptors, for which the conjugated addition of the Lewis base to the acid-activated monomer is the commonly accepted initiation mechanism. This has also been reported for B–P-based bridged Lewis pairs (BLPs) that form macrocyclic addition products.[1] In a detailed study of a series of Al–P-based BLPs, using a combination of single-crystal diffraction experiments (X-ray and neutron) - including an in-situ monitoring of a crystal to crystal transformation – and other methods an active role of the methylene bridge was revealed, acting as a base towards the α-acidic monomers.[2]

[1] T. Xu, E. Y.-X. Chen*, J. Am. Chem. Soc. 2014, 136,1774–1777.
[2] M. Weger, R. K. Grötsch, M. G. Knaus, M. M. Giuman, D. C. Mayer, P. J. Altmann, E. Mossou, B. Dittrich, A. Pöthig*, B. Rieger*, Angew. Chem. Int. Ed. 2019, 58, 9797.

S11-4

Combined spectroscopic and (photo)crystallographic studies of selected photo-excited multicentre coinage metal complexes

P. Łaski[1], J. Drapała[2], K. Durka[2], R. Kamiński[1], K. Jarzembska[1]
[1]University of Warsaw, Department of Chemistry, Warsaw, Poland
[2]Warsaw Institute of Technology, Department of Chemistry, Warsaw, Poland

Modern photocrystallographic methods allow us to inspect photo-induced short-lived excited-state species in crystal structures of photoactive compounds. Complexes of XI group transition metals at their 1st oxidation state are excellent examples of such compounds. Due to their specific electron configuration, copper(I), silver(I) and gold(I) complexes exhibit interesting optical properties which could be useful in the fields of materials engineering and optoelectronics. These properties could be further adjusted e.g. by the choice of organic ligands attached to the central metal atoms.

In this work, two model photoactive complexes have been thoroughly studied: a Ag(I)/Cu(I) tetranuclear complex and a Ag(I) mononuclear complex. Time-resolved photoluminescence spectroscopy combined with quantum computations have been used to characterise electronic transitions in the examined systems, and to determine lifetimes of the emissive excited states. Both samples appeared to be luminescent when irradiated with the 355 nm laser light. Emission maxima were located at 510 nm and 585 nm, respectively, while the emission decay times were in the micro second regime. Additionally, time-resolved X-ray pump/laser-probe Laue experiments, conducted at the APS synchrotron, have been performed in order to determine structural changes occurring along with the laser-light-induced excitations. The outcomes have been compared with the QM/MM computational results and shall be presented.

The authors thank NSC (2016/21/D/ST4/03753, 2014/15/D/ST4/02856) and WCSS (grant No. 285) in Poland, EU programme (POIG.02.01.00-14-122/09) and APS, USA (DOE: DE-AC02-06CH11357, NIH: R24GM111072) for financial support and access to facilities.

S11-5

Nitro – nitrito linkage isomerisation reaction in crystals of trinitrocobalt (III) organometallic complexes

K. Deresz[1], S. Kutniewska[1], R. Kamiński[1], A. Krówczyński[1], K. Jarzembska[1]
[1]University of Warsaw, Department of Chemistry, Warsaw, Poland

Nitro-cobalt complexes are known to have diverse applications, for example being already successfully used as catalysts in oxidation reactions[1] or conglomerate components[2]. As reported in literature, some of these complexes exhibit also thermoswitchable properties[3]. There is, however, limited information on light-induced linkage isomerism of this group of compounds.

In view of the above, a series of trinitrocobalt (III) complexes with different chelating amine ligands was chosen and thoroughly analysed. The idea of the project was to determine the optimal conditions of UV-vis light irradiation and temperature for inducing the nitro-group isomerisation reaction in the studied crystal structures. Furthermore, the impact of crystal packing and intermolecular interactions on the nitro group switching was studied. The chosen systems were investigated via multi-temperature (photo)crystallographic, spectroscopic and computational methods.

It appeared that temperature changes induce the isomerisation reaction in the studied crystal structures only to a small extent. On the other hand, the compounds may undergo the transformation to the (η2-O,O") –κ nitrito isomer by light irradiation in the wavelength range from 365 nm to 660 nm. The nitro groups engaged in relatively strong hydrogen bonds with the amine fragments of the adjacent molecules are not active, while the weaker-bound ones convert in about 25%. It is worth mentioning that the nitrito form is stable up about 240 K.

The PRELUDIUM grant (2017/25/N/ST4/02440) and the European Regional Development Fund (POIG.02.01.00-14-122/09) are gratefully acknowledged for financial support.

[1] B.S. Tovrog, S.E. Diamond, F. Mares, *Journal of the American Chemical Society*, 1979, 101, 270-272
[2] I. Bernal, *Inorganica Chimica Acta*, 1985, 101, 175-184
[3] T-H.T. Li, J. Breen, J.J. Worrel, *Journal of Coordination Chemistry*, 2009, 26, 15-34

S11-6

Controlled crystallization from deuterated solvents: Notable insights on crystallization behavior of brucine-multicomponent systems

A. Białońska[1], K. Merz[2], M. Yanbaeva[2]
[1]Wrocław University of Science and Technology, Faculty of Chemistry, Wrocław, Poland
[2]Ruhr University Bochum, Department of Inorganic Chemistry I, Bochum, Germany

Introduction

H/D-exchange as the smallest possible element substitution in the molecular structure affects the physical properties that depend directly on the proton or deuterium. In general, it is expected that the crystal structures of compounds containing hydrogen are not affected by the substitution of hydrogen by deuterium. A detailed survey of deuterated and non-deuterated compounds indicates that the assumption is wrong. Isotopic substitution can influence the molecular arrangement in the solid state. Particularly striking is the fact that an isotopic effect can also be observed in the use of solvents in crystallization procedures. Deuterated solvents can control the crystallization in selected cases of small organic compounds.

Objectives
Brucine, *N*-(3-nitrobenzoyl)aspartic acid, diastereomeric salts

Materials and methods
Crystallization studies

Results
Racemic resolution of *N*-(3-nitrobenzoyl)aspartic acid by formation and fractional crystallization of brucinium diastereomeric salts is the example in which solvent plays the crucial role. The racemic resolution is effective when crystallization is performed from methanol solution and ineffectual from solution containing water. In the later case depending on the molar ratio of the resolving agent to the resolved compound, the solid solution or the double salt precipitates. Applying deuterated solvents for the separation of the aspartic acid derivative by formation of brucinium salts leads to the solid solution precipitation.

Conclusion
In conclusion, using the example of brucine-multicomponent systems, we were able to show that the aggregation of molecules in the solid state is highly susceptible to small changes in the nature of the initial molecules. Deuterated solvents can be used for the targeted control of crystallization processes.

Merz, A. Kupka, *Cryst Growth Des*, 2015, *15*, 1553
D. D. Enkelmann, D. W. M. Hofmann, K. Merz, *Cryst. Growth Des* 2017, 17, 4726.
J. Falk, D. W. M. Hofmann, K. Merz, *IUCrJ*, 2018, 21, 2130.

Inorganic crystal structures II

S12-1
Crystal symmetry aspects of materials with magnetic spin reorientation

R. Przenioslo[1], P. Fabrykiewicz[1], I. Sosnowska[1]
[1]University of Warsaw, Department of Physics, Warsaw, Poland

Introduction
Spin reorientation is an important phenomenon which occurs in a number of oxides, e.g. haematite, alpha-Fe2O3 [1] rare earth orthoferrites [2] and intermetallics. Most microscopic models of spin reorientation describe either a sudden spin-jump or a continuous rotation of the magnetization direction within some plane of the crystal lattice [3].

Objectives: The goal is to introduce the information about the magnetic ordering symmetry [4-6] to the phenomenological description of the continuous spin-reorientation [3].

Materials and methods: High resolution synchrotron radiation measurements were performed for alpha-Fe2O3 [7]. The Tables of magnetic space groups [6] has been searched in order to find those groups which allow a continuous spin reorientation.

Results: It is shown that continuous spin reorientation is possible only for a few magnetic space groups and a few Wyckoff positions [8]. A list of these groups and positions is given [8]. A literature search for crystal structure studies of materials with spin reorientation has been done [8].

Conclusions: The crystal structure of materials with a continuous spin reorientation can be only triclinic or monoclinic i.e. it cannot be orthorhombic, tetragonal, trigonal, hexagonal or cubic [8]. This conclusion is confirmed for the case of the monoclinic symmetry of alpha-Fe2O3 [7].

References
[1] J. Morin, Phys. Rev. 78, 819–820 (1950).
[2] Sosnowska, E. Steichele, & A. Hewat, Physica B+C, 136, 394 (1986).
[3] M. Levinson, M. Luban, & S. Shtrikman, Phys. Rev. 187, 715 (1969).
[4] V. Shubnikov, "Simmietria i Antisimmietria Konechnych Figur", Moscow (1951)
[5] Opechowski & R. Guccione, Magnetism, Vol. 2A, edited by G. T. Rado & H. Suhl. New York: Academic Press (1965).
[6] B. Litvin, Acta Cryst. A64, 419 (2008).
[7] Fabrykiewicz, R. Przenioslo, I. Sosnowska and F. Fauth, Acta Cryst B74, 660 (2018).
[8] Przenioslo, P. Fabrykiewicz and I. Sosnowska, Acta Cryst A74, 705 (2018).

S12-2
$MnBi_2Te_4 \cdot (Bi_2Te_3)m$ (m=0÷6) compounds as candidates for magnetic semiconductors with non-trivial topology: crystal structure and some physical properties

Y. N. Aliyeva[1], P. A. Askerova[1], I. R. Amiraslanov[2], N. T. Mamedov [1]
[1]Institute of Physics, Azerbaijan National Academy of Science, Baku, Azerbaijan
[2]Institute of Physics, Azerbaijan National Academy of Science, Lab. of Crystallography, Baku, Azerbaijan

Topological insulators (TI) have come to the fore of the frontier research at the beginning of XXI century. Since then different types of non-trivial topologies in non-magnetic materials have been disclosed both theoretically and experimentally. However, magnetic materials have not been addressed for a long time since magnetic field breaks time reversal symmetry that is absolutely essential for existence of the non-trivial topology of a material. Nevertheless, a deep insight into the problem made in 2010 has shown that some antiferromagnetic (AFM) materials can support topological surface states which, along with exotic physics, are of particular interest for spintronics and quantum computing.

Since then a great deal of efforts have been made to obtain AFM TI by doping a 3D van der Waals topological insulator Bi_2Te_3 and its analogies with magnetic transition metal atoms. However the results were rather ambiguous. At last, in 2018 the AFM topological state was predicted and solidly verified on stoichiometric $MnBi_2Te_4$ /1/. Here we present the results of our studies on phase formation in MnTe-Bi_2Te_3 system. A series of homologous compounds of $MnBi_2Te_4 \cdot (Bi_2Te_3)m$ - type , where m=0÷6 were found and their crystallographic characteristics, as well as some physical properties were determined.

/1/. M. Otrokov *et al.* **arXiv:1809.07389 [cond-mat.mtrl-sci] (Nature *in press*)**

S12-3
Phase transition and magnetism in the synthetic mineral Fe3(PO3OH)4(H2O)4

M. Gutmann[1], M. Poienar[2], L. Pascut[3], G. Stenning[1], C. Paulmann[4,5], M. Tolkiehn[4]
[1]Rutherford Appleton Laboratory, ISIS Facility, Chilton Didcot, United Kingdom
[2]National Institute for Research and Development in Electrochemistry and Condensed Matter, Timisoara, Romania
[3]Stefan Cel Mare University, MANSiD Research Center and Faculty of Forestry, Suceava, Romania
[4]Deutsches Elektronen Synchrotron (DESY), Beamline P24, PETRA III, Hamburg, Germany
[5]University of Hamburg, Mineralogisch-Petrographisches Institut, Ham burg, Germany

Iron hydroxyl phosphates are minerals with rich crystal chemistry and potential applications in life sciences, where they can be used as effective catalysts, electrode materials, etc. Structural and magnetic characterization has been performed for the Fe3(PO3OH)4(H2O)4 synthetic mineral in the 2K - 470K temperature range. A monoclinic isostructural phase transition has been found at ~211K, which has not been reported before. The isostructural phase transition was confirmed by specific heat measurements. In addition, susceptibility measurements reveal that antiferromagnetic ordering sets in at 12K. The recently solved low temperature crystal structure, implies a complex magnetic behavior. Possible mechanisms of the phase transitions will be discussed in terms of ab-initio methods and phenomenological bond-valence model. Synchrotron X-ray diffraction has been performed at P24 beamline within DESY laboratory. Specific heat, susceptibility measurements and additional laboratory X-ray diffraction, were performed at Materials Characterization Laboratory at ISIS within Rutherford Appleton Laboratory.

S12-4

Nitridosilicatephosphates by high-pressure / high-temperature synthesis

W. Schnick[1], O. Oeckler[2], L. Eisenburger[1]
[1]Ludwig-Maximilians-University Munich, Department of Chemistry, Munich, Germany
[2]Leipzig University, IMKM, Leipzig, Germany

Nitridosilicates and nitridophosphates have attracted much scientific and industrial interest over the last years, especially the use of (oxo)nitridosilicate-based luminescent materials in pc-LEDs (phosphor-converted light emitting diodes) and several reports on new and unsual nitridophosphates.[1, 2, 3] However, anionic networks consisting of both PN_4 and SiN_4 polyhedra were so far unknown despite their potential for structural diversity by silicon either being coordinated by four (SiN_4-tetrahedra) or even six N atoms (SiN_6-octahedra comparable to SiO_6 units in SiP_2O_7), as well as by the possibility of phosphorus being coordinated four-, five- or sixfold.[2,3] The compounds $EASiP_3N_7$ (*EA* = Sr, Ba) were synthesized by high-pressure / high-temperature synthesis (8 GPa, 1400 – 1700 °C) employing a multianvil press with a Walker-type module.[4] The structures of both compounds are isotypic to the polymorphs of the mineral barylite ($BaBe_2Si_2O_7$) with $SrSiP_3N_7$ exhibiting the barylite-1*O* structure while $BaSiP_3N_7$ predominantly crystallizes in the barylite-2*O* structure.[5] The structures of both compounds were elucidated by single-crystal X-ray diffraction. The sum formulae derived from X-ray data are consistent with energy dispersive X-ray spectroscopy on single crystals. A mixed Si/P occupancy is very likely for one of the tetrahedrally coordinated sites, which is consistent with lattice-energy calculations.

[1] M. Zeuner, S. Pagano, W. Schnick, *Angew. Chem.-Int. Ed.* 2011, *50*, 7754.
[2] S. Vogel, M. Bykov, E. Bykova, S. Wendl, S. D. Kloss, A. Pakhomova, S. Chariton, E. Koemets, N. Dubrovinskaia, L. Dubrovinsky, W. Schnick, *Angew. Chem.-Int. Ed.* **2019**, *58*, 9060.
[3] S. D. Kloss, W. Schnick, *Angew. Chem.-Int. Ed.* 2019, *58*, 7933.
[4] H. Huppertz, *Z. Kristallogr.*, 2004, *219*, 330.
[5] S. Merlino, C. Biagioni, E. Bonaccorsi, N. V. Chukanov, I. V. Pekov, S. V. Krivovichev, V. N. Yakovenchuk, Armbruster, *Mineral. Mag.* 2015, *79*, 145.

S12-5

CRYSTAL PHASES OF InSe-GaSe SYSTEM

I. R. Amiraslanov[1], Ю Ю Azizova[2], Z. Jahangirli[2], Y. R. Aliyeva[1], Y. N. Aliyeva[2], S. A. Nabieva[2]
[1]Institute of Physics, Azerbaijan National Academy of Science, Lab. of Crystallography, Baku, Azerbaijan
[2]Institute of Physics, Azerbaijan National Academy of Science, Baku, Azerbaijan

The semiconductor compounds GaSe and InSe are of great practical interest as promising materials for optoelectronics. These compounds crystallize into a layered hexagonal structure with four atomic layers in a unit cell and chemical bonding between Ga-Ga and In-In atoms in GaSe and InSe, respectively. At the same time the nature of interactions between In and Ga atoms in $Ga_xIn_{1-x}Se_2$ solid solutions is still to be unveiled in order to clarify whether Ga-In intermetallic bonds are formed. While $Ga^{3+}In^{1+}Se_2$ compound is known to crystallize into the tetragonal structure, the nature of phase transition between the hexagonal and tetragonal phases (H↔T), as well as composition x at which this transition occurs is yet to be clarified.

Above-stated called for studies of phase formation in GaSe – InSe system, which we then performed using X-ray diffraction, thermogravimetric and Raman spectroscopy methods.

It was found that, despite the fact that InSe and GaSe are isostructural compounds, a continuous range of solid solutions between terminate binary compounds do not exist. A diffraction peak irrelevant to the diffraction pattern of a solid solution of isostructural GaSe and InSe appears and is believed to be a manifestation of the ordered arrangement of Ga and In atoms due to formation of the ternary InGaSe2 compound. The region of existence of this ternary tetragonal modification is found. Stability of the InGaSe2 phase formation with changing x was also examined by using DFT calculations and comparing the total energies of the formation of a solid solution and a ternary compound. The formation of the latter was found to be preferable for intermediate x composition. The temperature and structural characteristics of phase transitions in $Ga_xIn_{1-x}Se_2$ were determined.

S12-6

Intercalation and structural study of layered $GaInS_3$ crystals with aminopyridine and ethylenediamine.

A. B. Ragimli[1], I. R. Amiraslanov[1], Y. R. Aliyeva[1]
[1]Institute of Physics, Azerbaijan National Academy of Science, Lab. of Crystallography, Baku, Azerbaijan

In condensed matter physics, intercalation of layered chalcogenides with ions leads to many interesting phenomena including charge density waves and two-dimensional superconductivity. At the same time the intercalation of above crystals with organic molecules has been relatively little studied, in spite of the interest to organic intercalates.

In this report, we present the interaction of crystals of various layered phases $Ga_{1-x}In_{1+x}S_3$ (x = 0 ÷ 0.5) with small organic molecules. The former mainly have a hexagonal close-packed structure, but one of these phases with the composition $Ga_{0.8}In_{1.2}S_3$ is characterized by an original rhombic structure. To obtain intercalates, we tested a number of organic molecules (4-Aminopyridine, p-Phenylenediamine, 4-Chloroaniline, Benzidine, 1.3-Diaminopropane, 4 - Ethylpyridine, Dietilentriamin, Pyrazine, Piperazine, Hydrazine, p-Anisidine, etc.), however satisfactory result was obtained only with 4-AP molecules. More precisely, as a result of direct interaction of rhombic and hexagonal crystals with 4-aminopyride (4-AP) molecules at a temperature of 200 ° C, a new intercalated compound with the composition Ga0.8In1.2S3 · 0.5(NH2–C5H4N) was obtained. Also investigated are the interactions of crystals previously intercalated with 4-AP with the above-mentioned organic molecules. It has been found that ethylenediamine replace 4-AP molecules. The reaction products were controlled mainly by x-ray and thermogravimetric methods.

Thus, $Ga_{1-x}In_{1+x}S3$ intercalates with 4AP and ethylenediamine molecules were obtained. It turned out that the process of intercalation of layered crystals with organic molecules is extremely selective.

Extreme/non-ambient conditions

S13-1
To what extent is the molecular symmetry lowering related to poly(a)morphism in a SnI_4 system?

K. Fuchizaki[1]
[1]Ehime University, Physics, Matsuyama, Japan

A molecular crystal SnI_4 exhibits the water-type polyamorphism; its crystalline phase, CP-I, stable at ambient conditions, transforms into two amorphous states, Am-II and Am-I, with different densities. The latter states have the respective stable liquid counterparts, Liq-II and Liq-I [1]. However, there is a critical difference from water; SnI_4 has no longer "no man"s land." Indeed, the density difference between Liq-II and Liq-I was shown to be ~0.4 g/cc [2].

Recently, we have found that the molecular symmetry lowers from T_d to C_{3v} upon the liquid-liquid transition (LLT) [3]. This finding then tempted us to examine whether the symmetry lowering commences in the CP-I field because the T_d-to-C_{3v} change is not incompatible with *Pa3* of the CP-I structure.

We conducted *in-situ* EXAFS measurement using a multianvil press, MAX-80, installed at KEK-AR, Japan. However, we could not detect any clear evidence on the molecular symmetry change at high temperatures, below which the melting curve of CP-I becomes abnormally flat against pressure. The situation remains unchanged when the system entered the high-pressure modification (CP-II) region on compression. An equation-of-state description could suitably quantify the incompressibility of a SnI_4 molecule [4].

In conclusion, we could identify no apparent lowering of the molecular symmetry in the CP-I field and a pressure region beyond the field. Therefore, the symmetry lowering is characteristic to the LLT. We can propose a consistent scenario on the CP-I-to-CP-II structural change, upon which the deformation of molecules is inevitable. We speculate that the deformation is essential in forming Am-I, whose structure remains unresolved.

[1] K. Fuchizaki *et al.*, J. Chem. Phys. **135**, 091101 (2011). [2] K. Fuchizaki *et al.*, J. Phys. Soc. Jpn. **82**, 033003 (2013). [3] K. Fuchizaki *et al.*, J. Chem. Phys. **150**, 114501 (2019). [4] H. Naruta, K. Fuchizaki, *et al.*, J. Phys.: Condens. Matter, in press.

S13-2
Reactions under pressure – how to force the reaction?

S. Sobczak[1], A. Katrusiak[1]
[1]Adam Mickiewicz University, Department of Chemistry, Poznań, Poland

Traditionally, the synthesis of new materials is dominated by solvothermal and solution-based methods. However, both these methods require considerable amounts of energy and often produce undesired byproducts. To overcome these problems, inspiration could be taken from nature. A vast majority of commonly used materials were made at high-pressure and high-temperature conditions in Earth's interior. Gigantic pressure and extreme temperature create a unique environment, for the efficient synthesis of stable, high-density materials. Although the pressure significantly influences the chemical equilibrium, this parameter is often neglected. To this day, a few pressure-induced chemical reactions have been thoroughly investigated.

Here we show how the unique thermodynamic conditions generated inside the diamond anvil chamber combined with the different liquid environment can be an excellent platform for inorganic and organic synthesis. Not only the pressure can allow for post-synthetic modification of metal-organic frameworks changing their physical and chemical properties, but also can led to a novel 3-dimensional porous materials stable at ambient conditions.[1,2] What is more, we have found that the influence of high-pressure can efficiently replace the strong basic catalyst used for disulphide metathesis, highly increasing the yield and selectivity of the reaction.[3]

This research was supported by funding from the Polish National Science Centre grant PRELUDIUM 2017/27/N/ST5/00693.

[1] Półrolniczak, A.; Sobczak, S.; Katrusiak, A. *Inorg. Chem.* 2018, *57* (15), 8942–8950.
[2] Sobczak, S.; Katrusiak, A.; S, S.; A, K.; Sobczak, S.; Katrusiak, A. *Inorg. Chem.* **2019**, *58* (17), 11773–11781.
[3] Sobczak, S.; Drożdż, W.; Lampronti, G. I.; Belenguer, A. M. A.; Katrusiak, A.; Stefankiewicz, A. R. *Chem. - A Eur. J.* 2018, *24* (35), 8769–8773.

S13-3
Phase relations and stabilities of $MgCO_3$

J. Binck[1], L. Bayarjargal[1], S. Lobanov[2], W. Morgenroth[1], V. Milman[3], B. Winkler[1]
[1]Goethe University Frankfurt, Institute of Geoscience, Frankfurt/Main, Germany
[2]Deutsches GeoForschungsZentrum GFZ, Section 4.3, Chemistry and Physics of Earth Materials, Potsdam, Germany
[3]BIOVIA Dassault Systemes, Cambridge, United Kingdom

During the subduction of oceanic crust significant amounts of carbon bearing sediments may be transported deep into the Earth's mantle, predominantly in the form of carbonates [1]. A possible coexistence of silicates and $MgCO_3$ was proposed for cold and oxidized oceanic slabs in the lowermost mantle [2]. Direct evidence for the presence of carbonates in the mantle is given by mantle derived diamonds that contain inclusions of $MgCO_3$ [3]. Above 80 GPa and 2000 K, model predictions and experiments have indicated that magnesite transforms into a new polymorph where carbon becomes tetrahedrally coordinated by oxygen and forms $\{C_3O_9\}_{6-}$ rings with sp_3 hybridized bonds [2,4,5]. However, alternative structures have also been proposed for the same *P,T*-conditions [6]. The experimental findings rely on the challenging interpretation of powder X-ray patterns only [2,4,5]. At much higher pressures (>115 GPa) and temperatures significantly exceeding 2000 K, even more uncertainty exists concerning the phase relations and stabilities of $MgCO_3$ polymorphs [2,4,6,7].

Here, we have studied the phase relations and stabilities of $MgCO_3$ by Raman spectroscopy in laser-heated diamond anvil cells at *P,T*-conditions covering those of the lower mantle and the core-mantle boundary, as well as by DFT-based model calculations. High quality Raman data obtained at pressures >1 Mbar in conjunction with the results of the modelling studies have allowed us to establish a phase diagram of $MgCO_3$ at lower mantle conditions thus clarifying previously inconsistent findings.

Financial support by the DFG (FOR2125/CarboPaT, BA4020) is gratefully acknowledged.

[1] Sleep & Zahnle, Geophys Res-Plan. 106, 1373 (2001)
[2] Maeda et al., Scien. Rep. 7, 40602 (2017).

[3] Wang et al., Earth Plan. Sci. Lett. 141, 293 (1996).
[4] Oganov et al., Earth Plan. Sci. Lett. 241, 95 (2008).
[5] Boulard et al., Nature Comm. 6, 6311 (2015).
[6] Pickard & Needs, Phys. Rev. B 91, 104101, (2015).
[7] Isshiki et al., Nature 427, 60, (2004).

S13-4

Structural diversity, phase transition and pores activation of Cd(II)-metal-organic frameworks based on 4,4-azopyridine and terephthalic acid

A. Półrolniczak[1], S. Sobczak[1], A. Katrusiak[1]
[1]Adam Mickiewicz University, Department of Chemistry, Poznań, Poland

One of the most important features of the metal-organic frameworks (MOFs) is their high surface area and tunable porosity. Due to that, MOFs are frequently used as a novel class of adsorbents, with capability for releasing guest molecules at specific conditions. MOFs already find their applications as extremally sensible sensors, sorbents of fuel gases, drug delivery platforms or storage materials. Nevertheless, the influence of the interaction between host framework and a guest located in the pores on the physical properties of these materials are not completely defined yet.

Herein, we present a MOF, namely abbreviated as AMU3, built from cadmium(II) cations linked by terephthalic acid and 4,4"-azopyridine. AMU3 crystallizes in orthorhombic space group *Cmce* and possesses a unique capability for changing his guests from DMF to acetonitrile. Moreover, the pores of AMU3 can be activated after immersing the bulk sample in CH3Cl or during extraction with a supercritical CO_2, resulting in an empty host structure. To investigate an influence of the different guest molecules on the mechanical properties of AMU3, we have designed two different experiments in which single crystal with DMF (AMU3@DMF) and acetonitrile (AMU3@CH3CN) occupying the pores, have been compressed up to 2.00 GPa. We have found that the isochoric compression of both, AMU3@DMF and AMU3@CH3CN induces a phase transition at 0.4 GPa, leading to monoclinic space group $P2_1/n$.

This work was supported by funding from the Polish National Science Centre (OPUS 10 No. UMO-2015/19/B/ST5/00262) and by grant POWR.03.02.00-00-I026/16 co-financed by the EU European Social Fund under the Operational Program Knowledge Education Development.

S13-5

Inducing metallophilic interactions in compressed crystals

M. Andrzejewski[1], S. Racioppi[2], P. Macchi[3]
[1]University of Bern, Department of Chemistry and Biochemistry, Bern, Switzerland
[2]Chalmers University of Technology, Göteborg, Sweden
[3]Politecnico di Milano, Milano, Italy

Introduction

Metallophilic interactions are weak interactions occurring in crystals containing M+ metal cations that have fully occupied d10 orbitals (e.g. Cu+, Ag+, Au+).1,2 Their energy is comparable with the energy of hydrogen bonds and depends mainly on the ligands attached to the cation and the anion. These interactions play a significant role in materials showing luminescence or conductive properties.

Objectives

The aim of research was to shorten the distance between adjacent metal cations, by means of a hydrostatic compression, inducing metallophilic interactions in order to generate electric conductivity in a crystal.

Materials and methods

Ag(I) and Cu(I) 5-(2-fluoro-4-pyridyl)tetrazolate coordination polymers were studied in a diamond anvil cell using Xray diffraction. The structural results were corroborated with periodic DFT calculations.

Results

As a part of the study, isostructural coordination polymers of Ag (I) and Cu (I) metal cations were compressed up to 10 GPa. Significant difference in the behavior of these compounds at high pressure was reported. Despite the fact of the isostructural relationship, solely the silver compound forms argentophilic interctions, whereas cuprophilic interactions below the sum of Van der Waals radii does not appear in the copper analogue. Although, the Ag···Ag contact contracted below the distance between two atoms in metallic silver, the compound did not exhibit electrical conductivity due to its structural features.

Conclusion

High-pressure may be an efficient tool to induce and enhance materials properties. These studies comprise the first example of metallophilic interactions induced by compression and on the different consequences of the pressure effect on isostructural complexes of two coinage metals.

References

[1]M. Jansen, Angew. Chem. Int. Ed., 26, 1987, 1098-1110.
[2]H. Schmidbaur, A. Schier, Angew. Chem. Int. Ed.,54, 2015, 746-784.

Quantum crystallography

S14-1
Chemical bonding analysis based on routine X-ray diffraction experiments

F. Kleemiss[1], H. Yanai[2], S. Grabowsky[1]
[1]University of Bern, Department of Chemistry and Biochemistry, Bern, Switzerland
[2]Tokyo University of Pharmacy and Life Sciences, School of Pharmacy, Tokyo, Japan

The chemical bond, as a crucial concept for the chemist in understanding their observations, is still a precarious subject of chemistry. There is no operator or function that can unambiguously define what a chemical bond is. Theoretical chemistry provides a variety of descriptors that can be used to describe different kinds and situations of bonding. Their major limitation is the restriction to the quantum-mechanical wavefunction, which is not observable experimentally. The technique of X-Ray Wavefunction Refinement (XWR)[1] is a powerful link between the world of quantum chemistry and crystallography, as the experimentally observed structure factors can be fitted by molecular orbitals.

Using this approach, the experimentally restrained wavefunctions can be used to study chemical bonding in depths, e.g. ultimately polarized olefins in push-pull systems. The bonding situation changes from a delocalized double-bond to a charge-separated single-bond by exchange of substituents.[2] In a second series of substances, the formal double bond is observed to be longer than the neighboring single bond, contradicting the usual picture of bond-length bond-strength correlation (Figure). Three related compounds were measured on a home diffractometer and refined using XWR, which yielded wavefunctions for further complementary bonding analysis.

Figure: Lewis structure (bottom) and refined distances from XWR.

The resulting bonding descriptors allow a deepened understanding of the bonding situation in these compounds and prove that quantum crystallography is not only possible on home diffractometers, but chemical questions can be addressed within a normal, carefully measured routine dataset.

[1] M. Woińska, D. Jayatilaka, B. Dittrich, R. Flaig, P. Luger, K. Woźniak, P. M. Dominiak, S. Grabowsky, *ChemPhysChem* 2017, 18(23), 3334.
[2] H. Yanai, T. Suzuki, F. Kleemiss, H. Fukaya, Y. Dobashi, L. A. Malaspina, S. Grabowsky, T. Matsumoto, *Angew. Chem. Int. Ed.* 2019, *58*, 8839.

Figure 1

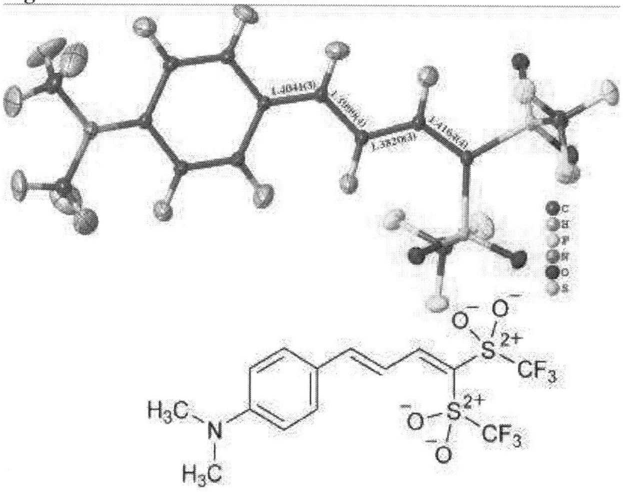

S14-2
Structure, the energetics of interactions, anharmonicity and reflection intensity cut-off effect: comparison of X-ray wavefunction refinement and multipole refinement for 2-hydroxy-8-*X*-quinoline derivatives (*X* = Cl, Br, I, S-Ph)

M. Woińska[1], M. Wanat[1], P. Taciak[2], T. Pawinski[2], W. Minor[3], K. Wozniak[1]
[1]University of Warsaw, Department of Chemistry, Warsaw, Poland
[2]Medicinal University of Warsaw, Warsaw, Poland
[3]University of Virginia, Department of Molecular Physiology and Biological Physics, Charlottesville, Virginia, United States

Quinoline derivatives are common among natural compounds and are utilized by industry in the synthesis of pharmaceutical compounds. X-ray crystallography is a great tool enabling investigations of the relationship between structure and interactions in potential drugs, given the progress in experimental techniques and the development of data processing methods. This work compares the commonly utilized method of high-resolution X-ray data refinement, the multipole refinement (MR), with the novel technique of X-ray wavefunction refinement (XWR).

Both methods were applied to data sets of varying quality, collected for the crystals of four quinoline derivatives substituted with Cl, Br, I atoms and the -S-Ph group. Two versions of refinement – including and excluding weak reflections ($|F|\geq 2\sigma(|F|)$) – were performed. The results obtained for the experimental structures were verified against the values calculated for structures optimized with dispersion-corrected periodic DFT. For the high-quality data sets (the Cl and -S-Ph compounds), both MR and XWR successfully refined the anisotropic displacement parameters and the positions of H atoms. The bond lengths obtained with XWR were more precise and closer to the theoretical values. In the case of the more challenging data sets (the Br and I compounds), only XWR enabled refinement of H atom geometrical parameters with results reflecting poor data quality. For both methods, the energy values (energy of intermolecular interactions, cohesive and geometrical relaxation energy) were in similar agreement with the values associated with the optimized structures with poor data quality being the source of the biggest divergencies. XWR allowed decreasing data-quality-related incorrect distortions of the reconstructed electron density. The attempt of refining anharmonic thermal motion (Fig. 1) revealed that weak reflections should not be rejected from the analysis, in order to achieve more correct interpretation of the obtained results.

Figure 1

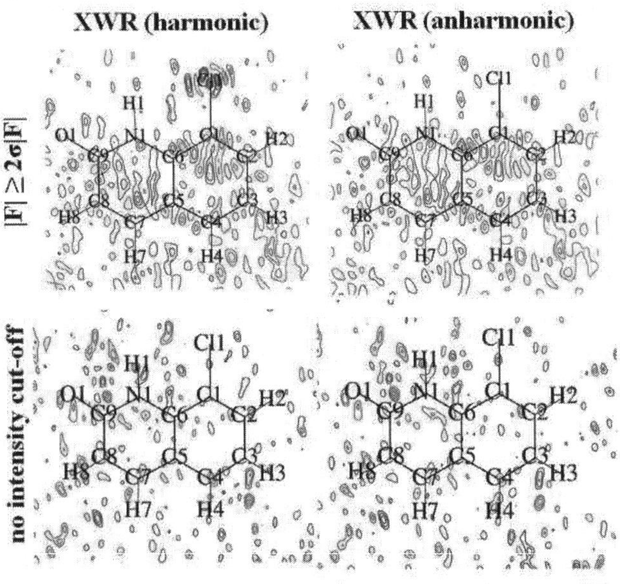

Fig. 1. PT-2(Cl) – residual density maps

S14-3
lamaGOET and the HAR-ELMO approach

L. A. Malaspina[1], E. K. Wieduwilt[2], A. Genoni[2], S. Grabowsky[3]
[1]University Bremen, Institute of Inorganic Chemistry and Crystallography, Bremen, Germany
[2]Université de Lorraine, CNRS, Laboratoire LPCT, Metz, France
[3]University of Bern, Department of Chemistry and Biochemistry, Bern, Swaziland

The use of quantum chemistry concepts and calculations in crystallographic refinements provides a new promising path for more accurate structure determination. One of such methods is Hirshfeld Atom Refinement (HAR)[1] which is performed and uniquely available within the program Tonto. Since HAR requires a quantum-mechanical calculation for each refinement step, it is computationally expensive and therefore it cannot be applied to larger systems such as macromolecules or compounds containing heavy elements. This problem is solved by coupling HAR with the newly developed databank of Extremely Localized Molecular Orbitals (ELMOs)[2]. This way, wavefunctions are not recalculated between the refinement steps, but constructed almost instantaneously from the deposited ELMOs.

In this work, we show that the accuracy and precision of hydrogen-atom parameters from HAR-ELMO refinements is as high as that from neutron-diffraction studies.[3] Subsequently, we discuss how to apply the new technique to polypeptides, proteins, but also coordination compounds with heavy elements. Based on the accurately derived geometries and ELMO wavefunctions, properties such as electrostatic potentials or electron localizabilities can be calculated as rapidly.

We also present the newly developed lamaGOET script, which provides a GUI for all the quantum crystallographic tools available in Tonto. lamaGOET interfaces Tonto with different software for the wavefunction calculation/construction, allowing the user to perform a HAR either by direct calculation of the wavefunction (using Gaussian, Orca or Tonto) or through transfer of ELMOs. lamaGOET is the only software available at the moment which is capable of performing HAR-ELMO refinements.[3]

[1] Capelli, S. C. *et al. IUCrJ* 2014, *1*, 361–379.
[2] Meyer, B. & Genoni, A. *J. Phys. Chem. A* 2018, *122*, 8965-8981
[3] Malaspina, L. A. *et al., J. Phys. Chem. Lett.* 2019, *10*, 6973-6982.

S14-4
Applications of HAR, TAAM and BODD methods for refinements against CuKα X-ray diffraction data

M. Wanat[1,2], M. Malińska[1], M. Gutmann[3], R. I. Cooper[4], K. Wozniak[1]
[1]University of Warsaw, Biological and Chemical Research Centre, Department of Chemistry, Warsaw, Poland
[2]University of Warsaw, College of Inter-Faculty Individual Studies in Mathematics and Natural Sciences (MISMaP) , Warsaw, Poland
[3]ISIS Neutron and Muon Source, Science and Technology Facilities Council, Rutherford Appleton Laboratory, Didcot, United Kingdom
[4]University of Oxford, Chemical Crystallography, Chemistry Research Laboratory, Oxford, United Kingdom

Introduction
Commonly, the Independent Atom Model (IAM) of electron density is used in the case of routine X-ray data analysis. However, this model does not give a quantitative description of electron density distribution. A far better model that allows for modelling of deformation of spherical charge density was introduced by Hansen and Coppens[1] and is called a pseudoatom model of electron density. However, application of this model requires an excellent quality of crystals and high resolution XRD data. Quite often, this is difficult to be fulfilled.

Objectives
The aim of the project is using new methods which have been developed for reconstruction of electron density against CuKα X-ray diffraction data. These methods are Bond-Oriented Deformation Density Model (BODD)[2], Hirshfeld Atom Refinement (HAR)[3] and Transferable Aspherical Atom Model (TAAM)[4].

Material and methods
CuKα and MoKα data of three model compounds were refined using IAM, BODD, HAR and TAAM methods. Final results were compared to the model obtained from neutron diffraction experiments.

Results
Results obtained with BODD, TAAM and HAR methods are contaminated by systematic errors (geometrical, atomic displacement parameters), however systematic errors of IAM are by far more serious.

Conclusion
BODD, TAAM and HAR refinements improve the CuKα XRD models.

Acknowledgments
This work was supported by the Foundation for Polish Science, TEAM-TECH Core Facility for crystallographic and biophysical research to support the development of medicinal products (co-financed by the European Union under the Regional Development Fund); Polish National Science Centre, MAESTRO grant No. DEC-2012/04/A/ST5/00609, ETIUDA grant No. UMO 2019/32/T/ST4/00086.

References

[1] Hansen, N. K.; Coppens, P. Acta Crystallogr. A 1978, 34 (6), 909–921

[2] Lübben, J. et al. Acta Crystallogr. A 2019, 75 (1), 50-62
[3] Capelli, S. C. et al. IUCrJ 2014, 1 (5), 361–379
[4] Jarzembska, K. N.; Dominiak, P. M. Acta Crystallogr. A 2012, 68 (1), 139–147

S14-5

THE QUANTUM CHEMICAL MODELING IN ORGANIC CRYSTALS STUDY

S. Shishkina[1], V. V. Dyakonenko[1], I. S. Konovalova[1], Y. A. Vaksler[1]
[1]V.N.Karazin Kharkiv National University, Inorganic Chemistry, Kharkiv, Germany

The use of quantum chemical methods to the study some features of molecular structure and crystal packing of organic compounds will be presented. Being powerful tool for modeling of conformations, different types of spectra, quantum chemical calculations need special attention in application to compare with X-ray diffraction data. First of all it concerns the calculations of molecules contained a π-conjugated system. A very soft delocalized π-electron system depends on polarizing environment and intermolecular interactions very much. So the calculations of molecular geometry in vacuo give a conformation that is very far from the experimental data. To approximate the calculations to the crystal phase geometry different levels of polarizing environment modeling should be used.

Another sphere of quantum chemical methods applying is the study of crystal packing. Our scientific group has developed a new approach to the analysis of crystal structure organization. It is based on the comparison of pairwise interactions energies calculated by *ab initio* quantum chemical methods. Application of this method to the study of a series of organic crystals allows us to suggest the classification of molecular crystal structures according to the type of basic structural motifs (BSM). All crystals may be divided onto four types:

1. i) crystals with isotropic packing of the building units;
2. ii) columnar crystals where BSM is chain/column;

iii) layered crystals with layers as BSM;

1. iv) columnar-layered crystals containing chains/columns as primary basic structural motif and layers as secondary BSM.

The application of the presented method of crystal structure analysis opens a way to study:

1. a) the role of different types of intermolecular interactions in crystal formation;
2. b) principles of crystal structure formation in different conditions of crystallization process;
3. c) anisotropy of crystal properties;
4. d) polymorphic transitions under pressure.

S14-6

Data processing and Hirshfeld atom refinements for an organo-gold(I) compound

S. Pawlędzio[1], M. Malińska[1], M. Woińska[1], K. Woźniak[1]
[1]University of Warsaw, Department of Chemistry, Warsaw, Poland

Relativistic effects in chemistry manifest themselves in many ways and influence various physical and chemical properties of materials. The well-known of them are the yellow color of gold or the high voltage of the lead-acid car battery[1,2]. Therefore, a description of these effects is of great importance for better understanding of the chemistry of heavy atoms.

A perspective method is quantum crystallography that relies on the high-resolution and high-quality XRD data to describe crystal structure in unprecedented detail. Intensities of the diffracted beam are affected not only by relativistic effects but also by absorption, anharmonic motion, anomalous dispersion[3], and others, which highly influence electron density distribution in the crystal and in consequence derived properties.

We collected the data sets for the chloro(triphenylphosphine)gold(I) using both Mo and Ag X-ray sources at 100K We also plan to perform X-ray diffraction experiments at a wide range of temperature (20 – 100K) using synchrotron radiation, which should allow us to distinguish anharmonic and bonding effects[4].

Here, we present the results of data processing and Hirshfeld atom refinements carried out for different resolution cutoffs. The outcome of DFT-based refinements with the nonrelativistic and quasi-relativistic approaches will be compared, including refinement of anharmonic thermal motions of the gold atom.

Support of this work by the National Science Centre, Poland through grant PRELUDIUM no. UMO-2018/31/N/ST4/02141 is gratefully acknowledged

[1] P. Pyykko, *Chemical Reviews*, 1988, 88, 563–594.
[2] K. B. Yatsimirskii, *Theoretical and Experimental Chemistry*, 1995, 31, 153–168.
[3] L. Bučinský, D. Jayatilaka and S. Grabowsky, *The Journal of Physical Chemistry A*, 2016, 120, 6650–6669.
[4] K. N. Jarzembska, K. Ślepokura, R. Kamiński, M. J. Gutmann, P. M. Dominiak and K. Woźniak, *Acta Cryst B*, 2017, 73, 550–564

Disordered Materials and Complex and aperiodic structures

S15-1
A disordered superspace approach to understand highly structured diffuse scattering

R. Neder[1], E. Schmidt[1]
[1]Friedrich-Alexander-University Erlangen-Nürnberg, Institute for Crystallography and Structural Physics, Erlangen, Germany

Single crystal diffuse scattering is generally interpreted using correlation parameters, that describe probabilities for certain configurations on a local scale. If the diffuse maxima are at a general position in reciprocal space many parameters are needed to simulate a short range ordered structure in direct space, which reproduces the observed diffuse maxima.

In the field of incommensurate crystallography a (3+d)-dimensional approach is taken to describe satellite reflections in reciprocal space, that cannot be indexed with integer (h,k,l) . A 3-dimensional aperiodic crystal structure is periodic in (3+d)-dimensional superspace and the atomic positions and/or occupancies of the 3D structure are described by modulation functions.

A perfectly periodic superspace gives rise to sharp satellite reflections. In order to describe boradend satellite reflections we introduce disorder into superspace. By introducing phase domains in the superspace structure we can generate structures that give rise to diffuse maxmia at any position in reciprocal space with an arbitrary width [1]. The disordered superspace approach also allows for a straight forward computational generation of a disordered structure using only few parameters.

The compound ThAsSe shows highly structured diffuse planes at $\mathbf{G}\pm0.14\langle110\rangle^{*}\pm\varepsilon\langle110\rangle^{*}\pm\eta[001]^{*}$ with ε and η essentially continuous [2]. The observed extinction conditions and the sharp diffuse rods can be directly interpreted using the disordered superspace approach. The diffuse planes in reciprocal are reproduced from a computational model crystal, that was build using the disordered superspace approach.

[1] Schmidt, E.M. and Neder, R.B.Phys.Rev. B 100, 054201.
[2] Withers, R.L. et al. Solid State Chemistry, 177 (2004) 701-708.

S15-2
A method for the quantification of stacking faults in the structures of NCA- and NCM-precursors for battery materials

S. Bette[1], B. Hinrichsen[2], R. E. Dinnebier[1]
[1]Max-Planck-Institute for Solid State Research, Scientific Facility X-ray diffraction, Stuttgart, Germany
[2]BASF SE, RAA/OS - M300, Ludwigshafen, Germany

Introduction
Li-bearing transition metal oxides, $Li(Ni_{1-x-y}Co_xAl_y)O_2$ (LNCA or NCA) or $Li(Ni_{1-x-y}Co_xMn_y)O_2$ (LNCM or NCM) are important constituents in state of the art Li-ion batteries. They can be obtained by the calcination of a mixture of a lithium salt like $LiOH{\cdot}H_2O$ and a Co- and Al-bearing nickel hydroxide as the NCA-precursor material. The NCA- and NCM-precursor materials are layered double hydroxide (LDH) phases that are structurally closely related to the brucite-type β-$Ni(OH)_2$. Due to the partial substitution of nickel in the cation sublattice, the ideal brucite-type structure is partially modified and defects occur as stacking faults and random intercalation of water or carbonate.

Objectives
In this study a method for the quantification of defects in the industrially synthesized precursor material is developed for the use as routine analysis.

Results
The microstructures, i.e. the fault probabilities (Fig. 1 a) were analyzed using arecursive routine for generating and averaging supercells of stacking faulted layered substances implemented into the TOPAS software. Automated series of simulations and refinements were performed in which the fault probabilities were modified incrementally and thus the microstructures optimized by a grid search (Fig 1 b, c). In total the amount of faulting was quantified in 15 different industrially synthesized precursors.

Conclusions
The amount of stacking faults in the materials corresponds to the degree of cation substitution in the β-$Ni(OH)_2$ lattice. Different degrees of interstratification faulting that were determined, could be correlated to the thermal analyses. The NCM precursor materials were generally less faulted than the NCA-precursor material.

Fig. 1. a) Illustration of the microstructure model applied for the NCA- and NCM-precursors, b) 3-dimensional grid of different fault probabilities, c) comparison of the fit using the faultless brucite-type structure and the most probable supercell model.

Figure 1

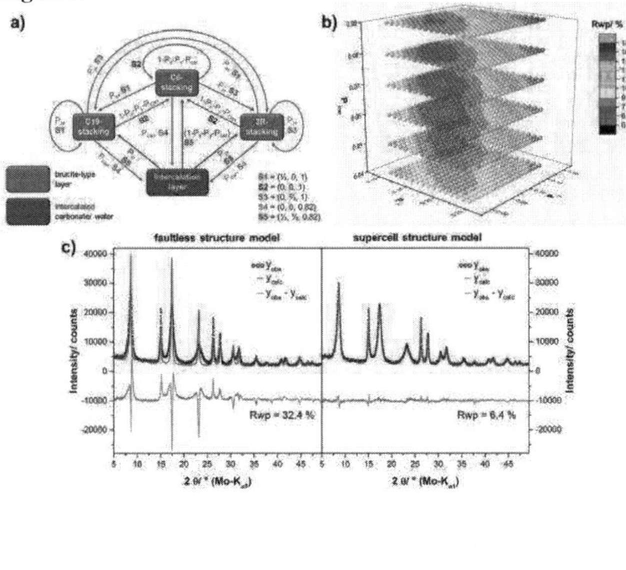

S15-3
RbSbO3 and (POCOP)CoBr: OD structures with two disorder modes

B. Stöger[1]
[1]TU Wien, X-Ray Centre, Vienna, Austria

The order-disorder (OD) theory[1] has been devised to explain the common occurrence of polytypism. OD structures are built of layers, which are related by partial operations (POs). If the POs are not valid for the whole structure, the stacking arrangement becomes ambiguous. Such an OD structure can crystallize in an infinity of polytypes, which are all locally equivalent up to at least pairs of adjacent layers. Therefore, the OD character of a structure may cause challenging crystallographic problems, such as twinning, anti-phase domains, allotwinning or disordered stacking arrangements.

Often, an OD structure is characterized by a layer contact with two possibilities of arranging the layers. Yet, in some cases there are more than one layer contact or adjacent layers can be arranged in more than two ways. Such structures may feature an even more complex crystallization behavior owing to a combination of e.g. twinning and disordered stacking. In this contribution two such structures are presented.

$RbSbO_3$ can be described as an alternating stacking of two types of trigonal layers, which each may appear in two orientations. Depending on crystallization conditions, allotwins (inter- or overgrowths of different polytypes) or single-polytype crystals are obtained. Additionally, different kinds of twinning are observed. These phenomena can be related to orientation change of either of the two layers.

The Co pincer complex (POCOP)CoBr is an OD structure of one kind of layer. Given a layer, the adjacent layer can appear in four positions, which are related by $m_{[100]}$ and/or $m_{[010]}$. Here likewise two kinds of stacking fault probabilities can be induced by variation of synthesis conditions. Crystals range from single domain to twins and disordered stackings with additional twinning.

[1] K. Dornberger-Schiff & H. Grell-Niemann. *Acta Cryst.* 14, 167–177 (1961)

S15-4

Enantiomeric disorder and modeling of X-ray diffuse scattering in copper(I) nitrate π-complex

D. Kowalska[1], V. Kinzhybalo[1], Y. I. Slyvka[2], M. Wołcyrz[1]
[1]Institute of Low Temperature and Structure Research, Polish Academy of Sciences, Wrocław, Poland
[2]Ivan Franko National University of Lviv, Department of Chemistry, Lviv, Ukraine

Studies concerning copper(I) compounds led to the novel crystalline π-coordination compound with 5-(allylthio)-1-(3,5-dimethylphenyl)-1*H*-tetrazole, [Cu(*m*-$C_{12}H_{14}N_4S)NO_3$]. The compound crystalizes in the orthorhombic space group $P2_12_12_1$ with cell parameters a = 7.94 Å, b = 12.76 Å, c = 14.71 Å. A trigonal pyramidal environment is arranged around Cu(I) ion. The organic ligand has a chelate function and nitrate plays a role of a bridging ligand, thus creating chains along the a axis. The chains, connected by hydrogen bonds and stacking interactions, form layers perpendicular to the c axis direction.

X-ray diffraction measurements reveal diffuse scattering streaks along c^* direction in the reciprocal space sections running through Bragg peaks. Only on the *0kl* reciprocal plane diffuse scattering streaks are extinguished (Fig. 1). This kind of phenomenon is being observed and described for the first time in this group of compounds. The arrangement of diffuse scattering streaks indicates on disorder existing among the atom layers perpendicular to the c axis.

For getting better insight in the local structure the modelling of diffuse scattering was performed. The structure refinement showed some extra electron density in the difference Fourier map in the places that would exactly match an enantiomeric form (mirror reflected) of [Cu(*m*-$C_{12}H_{14}N_4S)NO_3$]. Because there are strong correlations in the *ab* layers, only one of two possible chiral forms of [Cu(*m*-$C_{12}H_{14}N_4S)NO_3]_\infty$ chains could exist in a single layer (**P** or **E**) (Fig. 2). Thus models with disorder between atom layers, of **P** and **E** forms, were verified and the final model will be presented.

Fig. 1. *h2l* reciprocal layer with diffuse scattering streaks (there are no streak running through the center of the layer, for $h = 0$).

Fig. 2. *ab* – plane in primary (**P**, left) and enantiomeric (**E**, right) configuration. The unit cell (solid line) and the pseudo-symmetry planes (dashed lines) are shown.

Figure 1

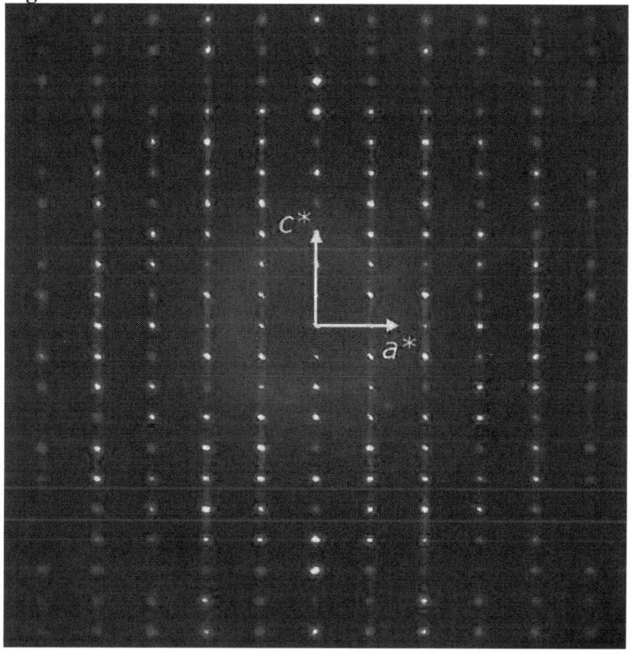

Figure 2

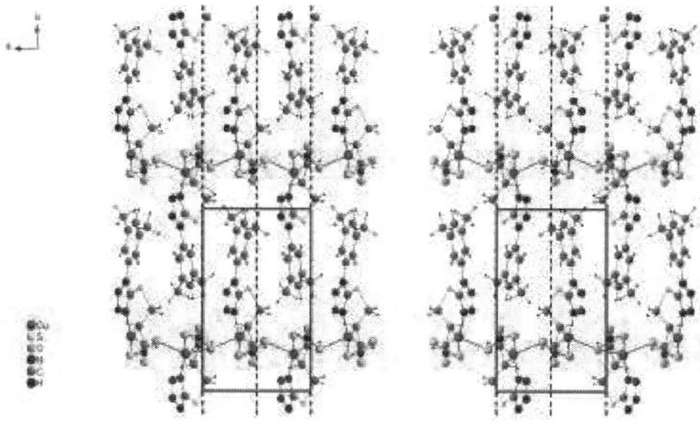

S15-5

Superspace structure of glycyl-L-valine

T. Rekis[1], S. van Smaalen[1]
[1]University of Bayreuth, Laboratory of Crystallography, Bayreuth, Germany

Glycyl-L-valine is a dipeptide that has been reported to crystallize in the $P2_12_12_1$ space group with the following unit cell parameters: $a = 5.52$ Å; $b = 26.6$ Å; $c = 44.1$ Å [1]. There are 7 molecules in the asymmetric unit (Z'=7). Structures with $Z' > 4$ are extremely rare and their formation is sometimes associated with translational modulations [2]. Such pseudo-symmetry is present also in the case of glycyl-L-valine. Inspection of the diffraction data reveals that the reported 3-dimensional supercell structure is preferably described as a 7-fold substructure in (3+1)-dimensional superspace with unit cell parameters a, b, and $c/7$. The structure is commensurate at the studied temperatures. Furthermore, the modulation wavevector is found to be $\mathbf{q}$ = (0, 2/7, 0) instead of (0, 1/7, 0) indicating that it is a genuinely modulated structure. Main stabilizing forces in the crystal structure of glycyl-L-valine are hydrogen bonds forming an extended network. Different parameters of those and their variation as well as molecule conformation variation along the superspace parameter t are discussed.

[1] Görbitz, C. H.; Augustsdottir, S.; Bleken, F. *Acta Cryst. C*, 2007, 63, o58 – o60
[2] Brock, C. P. *Acta Cryst. B*, 2016, 72, 807 – 821

S15-6

Multidimensional analysis *vs.* statistical method of diffraction and structure description of aperiodic systems

J. Wolny[1], R. Strzałka[1], I. Bugański[1], J. Śmietańska[1]
[1]AGH University of Science and Technology, Faculty of Physics and Applied Computer Science, Kraków, Poland

Over a hundred years of diffraction analysis for periodic crystals encountered serious difficulties when trying to describe diffraction images for aperiodic systems, such as modulated structures or quasicrystals. In order to restore periodicity in a direct space, a multi-dimensional approach is commonly used [1-3]. In this approach, the actual structure becomes an oblique view of the selected part of the multidimensional periodic structure. The multidimensional description defines both a periodic direct lattice and a symmetrically equivalent periodic reciprocal lattice, which allows the use of well-known methods of classical crystallography. The problem is the universality of this approach. For each case, we must redefine both the dimension of the space used and the so-called atomic surface, i.e. a mathematical object associated with a real atom, and describing the hypothetical position of the atom in additional dimensions. The problem is extremely complicated even for simple model mathematical structures. In addition, there are many model structures for which the structure does not become periodic in any space with a finite number of dimensions. We encounter even greater difficulties when describing the dynamics of multidimensional systems. Therefore, there was an attempt to describe any structure using only physical space (3D), giving up the notion of a direct and reciprocal lattice with associated symmetry elements. This method uses the Fourier transform properties and is based on the calculation of appropriate distribution of the probability of atom positions relative to the reference grids associated with a given scattering vector [3-5]. Ultimately, this leads to the expression on a structural factor in the form of a multimodal Fourier transform with probability distributions. The paper will present examples of the use of this new method to describe diffraction images derived from a single slit, through periodic and quasiperiodic crystals, including modulated structures, both commensurately and incommensurately [3,6].

[1] Yamamoto, Acta Cryst. A 52 (1996) 509.
[2] Steurer, S. Deloudi, *Crystallography of quasicrystals. Concepts, methods and structures.* Berlin: Springer; 2009.
[3] Wolny, I. Bugański, R. Strzałka, Cryst. Rev. 24 (2018) 22.
[4] Wolny, Philos. Mag. A 77 (1998) 395–412.
[5] Strzałka, I. Bugański, J. Wolny, Crystals 6 (2016) 104–123.
[6] R. Strzałka, I. Bugański, J. Śmietańska, J. Wolny, JPCS 2019 (accepted).

Spectroscopy

S16-1

Analysis of Ferroelectric Strontium Titanate Thin Films with Resonant X-ray Diffraction

M. Nentwich[1], C. Richter[2], M. Zschornak[3,4], T. Weigel[3], D. Novikov[5], D. C. Meyer[6]
[1]Technische Universität Bergakademie Freiberg, Institute of Experimental Physics, Freiberg, Germany
[2]Insitut für Kristallzüchtung, Berlin, Germany
[3]Technische Universität Bergakademie Freiberg, Institute of Experimental Physics, Freiberg, Germany
[4]Helmholtz-Zentrum Dresden-Rossendorf, Institute of Ion Beam Physics and Materials Research, Dresden, Germany
[5]Deutsches Elektronen Synchrotron (DESY), P23, Hamburg, Germany
[6]Technische Universität Bergakademie Freiberg, Institute of Experimental Physics, Freiberg, Germany

Ferroelectric materials possess a spontaneous polarisation that can be electrically switched between two different orientations. Compared to ferromagnetics, the domain walls are small, allowing memory cells of higher storage density. Thin films are of interest as they allow to increase the storage density further. Exemplarily, incipient ferroelectricity is found in strontium titanate (STO) at a low transition temperature. This ferroelectricity can be stabilized by a variety of external factors such as doping, strain, electric field, isotope substitution etc., even up to room temperature. Recently, we studied the effect of the electroformation of STO and the accompanied creation of the migration-induced field-stabilized polar (MFP) phase [1] extensively, i.e. with the newly developed approach of Resonantly Suppressed Diffraction (RSD) [2] at Beamline P23 of PETRA III. Because of the breakdown of Friedel's law under resonant conditions, RSD can be used to obtain information about the polarization state of polar materials. Namely by monitoring the Bragg intensity while scanning the energy of the incident beam through the absorption edge of strontium. Here, we transferred the methodology to ferroelectric STO thin films and measured the energy dependent intensities for different Bragg reflections. We used three different structure models to fit the experimental data: the AFD phase, a generalized MFP phase, and a structure showing the displacements of both phases. With these fits, we could confirm that the ferroelectric phase is strongly related to the MFP phase sharing the same structural symmetry and, thus, will exhibit similar properties and applications.

[1] J. Hanzig et al.: *Migration-induced field-stabilized polar phase in strontium titanate single crystals at room temperature*, Phys Rev B 88 024104 (2013)
[2] C. Richter et al: *Picometer polar displacements in strontium titanate determined by a new approach of resonant x-ray diffraction*, Nature Commun 9 178 (2018)

S16-2

Characterization of dynamic thermal displacements of Yttrium manganese

T. Weigel[1], C. Richter[2], M. Nentwich[1], M. Zschornak[1], D. C. Meyer[1]
[1]Technische Universität Bergakademie Freiberg, Institute of Experimental Physics, Freiberg, Germany
[2]Insitut für Kristallzüchtung, Berlin, Germany

Resonant Elastic X-ray Scattering (REXS) can be used for a detailed study of electronic structures, defect structures, magnetic properties or atomic displacements. The latter, is especially important for the characterization of structure-property-relations. A novel technique in the field of REXS called Resonantly Suppressed Diffraction (RSD) uses the suppression of allowed reflections to characterize displacements in the picometer range [1]. RSD tunes the photon energy such that certain reflections approach zero due to destructive interference. The method has been successfully applied to determine the structure of the recently discovered migration-induced field-stabilized polar (MFP) phase in SrTiO3 [2], where a determination of the crystal structure was not possible with conventional X-ray diffraction.

The sensitivity of RSD makes it the perfect probe to investigate the structure of the commensurately modulated (CM) low temperature phase in YMn2O5. This phase is ferroelectric, but it is not clear if the ferroelectricity origins from small atomic displacements or from the electronic spin configurations of Mn2+ ions. Here, we employ RSD on carefully chosen Bragg reflections and we extract the atomic displacement parameters (ADP) of the paraelectric phase, measured close to the phase transition at 50 K. We also calculated the RSD curves for the tested Bragg reflections with ADPs that we found in literature, all showing worse results than our fitted values.

The knowledge of these ADPs is the basis to clarify if the CM phase also exhibits a structural modification in addition to its magnetic modulation. With this study we want to show that RSD is a unique approach for the characterization of structural particularities, where other structure determination methods are not sensitive enough or not applicable at all.

[1] C. Richter et al.: Nature Communications 9, 178 (2018).
[2] J. Hanzig et al.: Physical Review B 88, 024104 (2013).

S16-3

Trivalent lanthanide and Actinide Incorporation into Zirconium(IV) oxide – Spectroscopic Investigations of Defect Fluorite Structures

M. Eibl[1], S. Shaw[2], C. Hennig[1], D. Prieur[1], K. Morris[2], T. Stumpf[1], N. Huittinen[1]
[1]Helmholtz-Zentrum Dresden-Rossendorf, Dresden, Germany
[2]Research Centre for Radwaste Disposal and Williamson Research Centre, School of Earth, Atmospheric and Environmental Science, Manchester, United Kingdom

Zirconia (ZrO_2) can incorporate a variety of metal cations with differing oxidation states up to high dopant loadings. In the case of trivalent metal ion incorporation, the ZrO_2 crystal structure may transform from the monoclinic (m) crystal phase to the tetragonal (t) or the cubic (c) ones depending on the overall amount of the trivalent dopant, with a subsequent introduction of oxygen vacancies into the lattice.[1]

In these investigations the incorporation of Eu^{3+} or Y^{3+} and the subsequent stabilization of the zirconia crystal structure have been studied both on the bulk level and the dopant level using a variety of experimental methods.

PXRD was used to study the ZrO_2 phase transformation. Luminescence spectroscopy (TRLFS) and X–ray absorption spectroscopy (EXAFS) were combined to study the local environment of Eu^{3+} or Y^{3+}/Eu^{3+} co–doped ZrO_2 samples. To bridge the gap between the bulk and the dopant-specific investigations, X-ray pair distribution function analysis (XPDF) was performed for selected Eu^{3+}–doped samples.

With increasing Eu^{3+} doping in ZrO_2, a transition from the monoclinic phase to the stabilized ones can be observed. The luminescence studies show the presence of multiple Eu^{3+} environments in m-ZrO_2. In the stabilized phases, a very constant luminescence signal is obtained, showing a maximum Stark splitting indicative of a low symmetry environment. Thus, the high symmetry of the bulk does not translate to the local site symmetry. Both XPDF and EXAFS studies confirm the very constant dopant environment over a broad range of doping percentages.

The presence of oxygen vacancies is assumed to make the ZrO_2 lattice flexible enough to accommodate the oversized trivalent cations. These vacancies are preferably located in the coordination sphere of the Zr ions, rather than around the dopant[2]. Thus, no change can be seen in the direct environment of the dopant with the spectroscopic techniques.

[1] Kelly, J., *Dent. Mater.* 2008 24, 289
[2] Stapper, G., *Phys. Rev. B* 1999 59, 797.

S16-4

Curium incorporation in rhabdophane solid solutions ($La_{1-x}Gd_xPO_4$•$0.67H_2O$)

N. Huittinen[1], A. Scheinost[1], Y. Ji[2], P. Kowalski[3], Y. Arinicheva[3], S. Neumeier[3]
[1]Helmholtz-Zentrum Dresden-Rossendorf, Dresden, Germany
[2]China Academy of Engineering Physics, Sichuan, China
[3]Research Center Jülich GmbH, Jülich, Germany

Introduction

Rhabdophane solids (*Ln*PO_4•$0.67H_2O$, *Ln* = La to Gd) are alteration phases of monazite ceramics (*Ln*PO_4), which have been considered as potential waste forms for the immobilization of specific radioactive waste streams containing actinide elements. In the event of monazite dissolution in underground repositories and the subsequent recrystallization of rhabdophane, incorporation of actinides within the secondary phase may occur.

Objectives

Very little is known about actinide incorporation within the rhabdophane structure. Therefore, the present study aims at a detailed understanding of the incorporation of curium (Cm^{3+}) in $La_{1-x}Gd_xPO_4$•$0.67H_2O$ rhabdophane solid solutions[1].

Materials and methods

Luminescence spectroscopy (TRLFS) was used to study the incorporation of Cm^{3+} in rhabdophane solids. To explain the spectroscopically observed substitution of Cm^{3+} for the various site types in the investigated solid phases, *ab initio* calculations were performed.

Results

The luminescence spectroscopic data shows that several Cm^{3+} species are present in the synthesized rhabdophane solids, depending on composition. The site occupancy derived from our TRLFS investigations show preferential incorporation of Cm^{3+} on the nonhydrated site. This is not in agreement with the fully hydrated rhabdophane structure where two-thirds of the sites are associated with water of hydration, leaving only one-third of the sites nonhydrated.

Conclusions

The preferential incorporation of Cm^{3+} on rhabdophane lattice sites that are not associated with water of hydration, imply that structural substitution reactions cannot be predicted based on the structure of the host matrix only. An important consequence of this favored incorporation of actinide dopants on specific lattice sites may be an unpredictable release of radioactive contaminants due to incongruent dissolution of the solid matrix.

[1] N. Huittinen et al. (2018) Inorg. Chem. 57, 6252–6265

Instrumentation

S17-1

X-ray and neutron single-crystal diffraction in diamond anvil cells

A. Grzechnik[1], M. Meven[1], K. Friese[2]
[1]RWTH Aachen University, Institute of Crystallography, Aachen, Germany
[2]Research Center Jülich GmbH, Jülich, Germany

Single-crystal X-ray diffraction in a diamond anvil cell (DAC) is routinely performed to megabar pressures. However, there are hardly any single-crystal neutron diffraction studies in the DAC that would present complete structural refinements. Even at the most advanced neutron facilities, it is difficult to study crystals with volumes below 1 mm3 due to the low flux of the neutron beams. The requirement for large samples hinders a joint use of X-ray and neutron single-crystal diffraction on compression. The combination of both techniques is highly advantageous for studies of crystalline compounds, as X-ray diffraction fails to provide precise information on, for instance, magnetic order or hydrogen bonding.

We have developed *transmission* cells suitable for both neutron and X-ray single-crystal diffraction with an opening angle of 80°. One of them could be operated remotely using a membrane filled with the He gas. The same crystals in the same cell can now be studied on the single-crystal diffractometer HEiDi at the Heinz Maier-Leibnitz Zentrum in Garching as well as on laboratory X-ray diffractometers and synchrotron beamlines [1]. We have now been working on procedures to combine the neutron and X-ray data for joint refinements of crystal structures using the program Jana2006 [2]. A newly developed DAC with larger opening angles than 80° and made of the Ni-Cr-Al alloy [3] allows for a wider access to the reciprocal space. We will present examples of structural refinements from X-ray and neutron data measured in our *transmission* DACs.

[1] A. Grzechnik, M. Meven, C. Paulmann, K. Friese, *J. Appl. Cryst.* (2019), accepted
[2] V. Petricek, M. Dusek, L. Palatinus, *Z. Krist.* (2014), 229, 345.
[3] Y. Cheng *et al.*, *Materials Science and Technology* (2019), https://doi.org/10.1080/02670836.2019.1578077.

S17-2

On the Design of a Dedicated Electron Diffractometer

E. Hovestreydt[1], G. Santiso-Quinones[1], G. Steinfeld[1]
[1]ELDICO Scientific, Villigen, Switzerland

www.eldico-scientific.com

Electron Diffraction (ED) as such has been around since the early days of Electron Microscopy. Since Transmission Electron Microscopes (TEMs) are available with accelerating powers of 200 to 300 kV and 2D detectors have become fast enough, Electron Crystallography really took off.

So far, ED has been done in TEMs that are modified, resulting in challenging experiments and limited datasets, yet, structures could be obtained from samples in the range of merely tens of nanometers, that were unsolvable with either conventional or even synchrotron X-ray radiation.

For some reason, no dedicated Electron Diffractometer has been available commercially so far. Data quality would greatly benefit from a TEM setup that focuses on the diffraction capability over imaging and allowing for faster and more complete datasets.

We will present a possible Electron Diffractometer design for Electron Crystallography from the point-of-view of X-ray Crystallography and indicate improvements over present TEM-based as well as X-ray instruments. It is envisioned, that preliminary results will also be presented.

"*Design guidelines for an electron diffractometer for structural chemistry and structural biology.*", J. Heidler et al., *Acta Cryst.*, 2019, *D75*, 458-466.
"*Rapid Structure Determination of Macrocrystalline Molecular Compounds using Electron Diffraction.*", T. Grüne et al., *Angew. Chem. Int. Ed.*, Oct 16, 2018, *57*, 16313-16317.

Fig. 1: ELDICO Electron Diffractometer

Figure 1

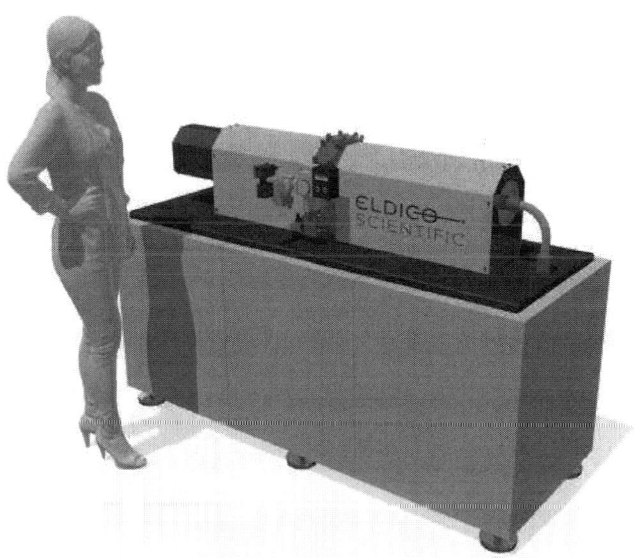

S17-3

Two new diffractometers at BM20/ESRF for single crystal, powder and surface diffraction

C. Hennig[1,2], M. Schmidt[1], A. Ikeda-Ohno[1,3], T. Radoske[1], M. Feig[4], S. Findeisen[1], J. Claussner[1], J. Exner[1,2], D. Naudet[1,2], N. Baumann[1,2], A. Scheinost[1,2]
[1]Helmholtz-Zentrum Dresden-Rossendorf , Institute of Resource Ecology, Dresden, Germany
[2]The Rossendorf Beamline at ESRF, Grenoble, France
[3]Japan Atomic Energy Agency, Tokai-mura, Japan
[4]Technische Universität Bergakademie Freiberg, Institute of Experimental Physics, Freiberg, Germany

The Institute of Resource Ecology / Helmholtz-Zentrum Dresden-Rossendorf is operating the Rossendorf Beamline (ROBL/BM20) at the European Synchrotron Radiation Facility (ESRF) for 20 years [1]. We are constructing a second experimental hutch with two new diffractometers for single crystal, powder and surface diffraction.

The single crystal diffractometer is foreseen for small and large molecule crystallography, mainly for structures with heavy metals. This diffractometer is equipped with a Pilatus3 X 2M detector mounted in a sturdy metal frame on a granite table. Sample-detector distances can be varied between 140 und 600mm. Three goniometers are available to mount single crystals: a Huber Kappa goniometer 512.410, an Arinax Kappa MK3, and a Huber uniaxial goniometer 410. The energy range of 5-35keV allows the application of anomalous dispersion. In-situ experiments on single crystals and powders are supported. Diffraction measurements can be combined simultaneously with XANES and XRF spectroscopy using a Vortex X90 CUBE silicon drift detector with a FalconX1 processor.

The second diffractometer will be used for surface diffraction and moderate high-resolution powder diffraction. This diffractometer is a 6-circle Huber diffractometer with Eulerian cradle geometry. Surface and powder diffraction techniques comprise specific setups. Both techniques will use a Pilatus 100k detector. The resolution for powder diffraction using the Pilatus 100k detector does not reach the resolution of synchrotron diffractometers with secondary analyzer crystals, but provides a magnitude better resolution than lab diffractometers. A specific support allows the installation of the Vortex X90 detector also on this diffractometer.

The setup comprises a cryo cooler (80-400 K) and a heater (up to 1200 K) which can be installed at both diffractometers. The experimental hutch is equipped for the use of radioactive material.

[1] http://www.esrf.eu/UsersAndScience/Experiments/CRG/BM20

Figure 1

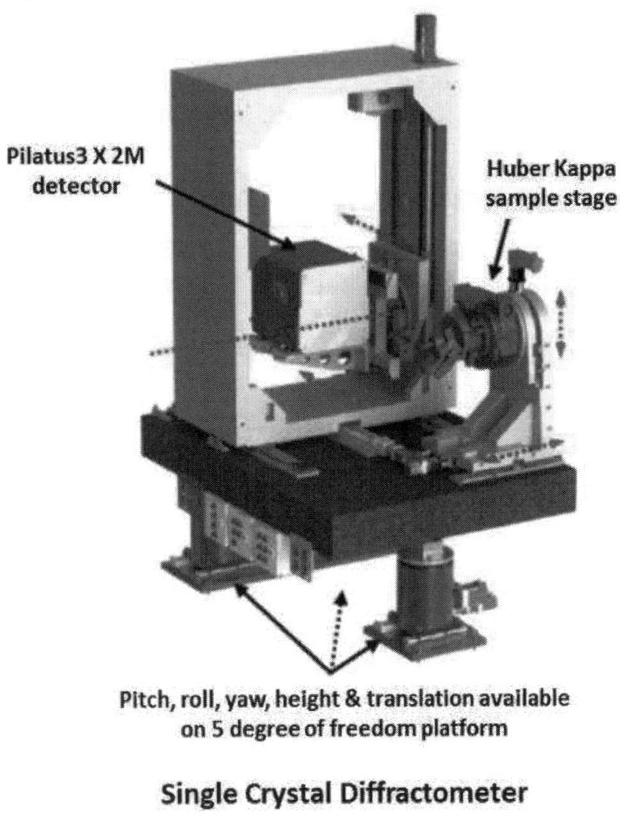

Figure 2

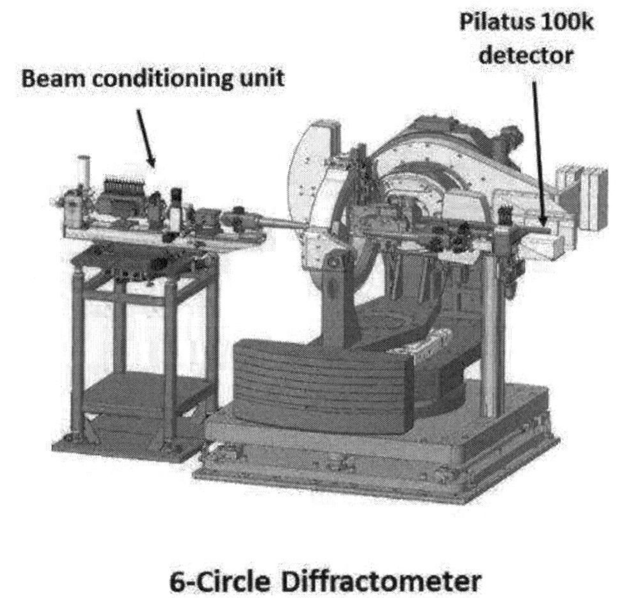

S17-4
The Benefits of Cu-Kβ Radiation

M. Bodensteiner[1], T. Mayr[1], F. Meurer[1]
[1]University of Regensburg, X-ray crystallography, Faculty of Chemistry and Pharmacy, Regensburg, Germany

The Cu-Kβ wavelength (λ = 1.39 Å) has been known for a very long time, mostly as result of an unsuccessful alignment of the diffractometer optics. However, there are benefits and powerful features of this radiation that many crystallographers are not aware of. Compared to Cu-Kα (λ = 1.54 Å) the amount of available data is increased by more than 35 per cent and the absorption significantly lowered. This is also true for a number of elements (Cu to Y) compared to Mo-Kα radiation (λ = 0.71 Å). In addition, the diffraction power as well as the detective quantum efficiency is much higher for Cu-Kβ relative to Mo-Kα. Another general advantage of Kβ radiation compared to Kα is the absence of α1/α2 splitting at higher diffraction angles. This leads to a relative improvement of the I/σ(I) at higher resolution. Our investigations have shown that in many cases almost identical or even better quality structures could be obtained by using Cu-Kβ compared to either Mo or Cu standard Kα wavelengths. The presentation will show this at a number of examples and provide insight into the technical and refinement procedures within Cu-Kβ experiments.

Figure 1

Experiment (Ylid $C_{11}H_{10}O_2S$)	Cu-Kα	Cu-Kβ	Mo-Kα
exposure time(s)	1s, 4s	1s, 4s	20s
a / Å	5.95813(10)	5.95788(10)	5.9813(3)
b / Å	9.03112(14)	9.03199(14)	9.0675(4)
c / Å	18.3817(3)	18.3822(3)	18.4452(10)
V / Å³	989.09(3)	989.17(3)	1000.39(9)
ρ(calc) g/cm³	1.385	1.385	1.369
μ / mm⁻¹	**2.658**	**1.944** (-27%)	**0.292** (-89%)
λ / Å	1.54184	1.39222	0.71073
2Θ range for data collection /°	9.624 to 148.638	8.688 to 149.064	6.302 to 64.276
Reflections collected	**7456**	**10575** (+42%)	**5282** (-29%)
Independent reflections	**2004**	**2749** (+37%)	**3179** (+59%)
R(int)	0.0177	0.0313	0.0188
$I/\sigma(I)$ to 0.84 Å	**39.45**	**17.13**	**21.55**
$I/\sigma(I)$ to max. resolution	**38.83** (0.80 Å)	**15.61** (0.72 Å)	**15.35** (0.67 Å)
Goodness-of-fit on F^2	1.145	1.076	1.074
Final R indexes $I>2\sigma(I)$	R_1 = 0.0289, wR_2 = 0.0710	R_1 = 0.0307, wR_2 = 0.0776	R_1 = 0.0388, wR_2 = 0.0875
Final R indexes all data	R_1 = 0.0290, wR_2 = 0.0710	R_1 = 0.0312, wR_2 = 0.0782	R_1 = 0.0443, wR_2 = 0.0917
Largest diff. peak/hole / e Å⁻³	0.19/-0.40	0.18/-0.44	0.24/-0.22
Flack parameter	-0.004(8)	-0.021(9)	0.03(4)

S17-5
IDEAL - What Is It Good For?

H. Ott[1], M. Ruf[2], J. Lübben[1], C. B. Hübschle[1]
[1]Bruker AXS GmbH, Application, Karlsruhe, Germany
[2]Bruker AXS Inc, Madison, United States

For more than 50 years crystallographers successfully apply the Independent Atom Model (IAM) using spherical scattering factors to model atoms in structures derived from X-ray data. However, the IAM over-simplifies the description of electron densities since the charge cloud of an atom is hardly ever spherical. The scientific field of charge density investigation better describes these phenomena by introducing elaborate multipole models. On the other hand, for routine structure investigations, the IAM was traditionally sufficient, as accessible data resolution was low and measurement time scarce.

This paradigm changes with Bruker"s advanced SC-XRD instrumentation: large photon-counting PHOTON III detectors and high brilliance IμS-type X-ray sources enable crystallographers to routinely collect high-precision, high-resolution data with short experiment times. Now, the newly developed Invariom (1) Derived Electron Analysis (IDEAL), the latest add-on to the APEX3 suite, intuitive and comprehensive tools to make full use of your superior data.

IDEAL goes beyond the traditional Independent Atom Model by expanding it to aspheric atom models using scattering contributions from bonds and lone pairs.(2) IDEAL improves structure models significantly making full use of the current data quality. IDEAL is easy-to-use within the APEX3 software suite and fully integrated into the IUCr"s checkCIF routines.

Due to its ease-of-use and significantly improved refinement results IDEAL sees ever-growing popularity.

We will report on the improvements of the model quality, which include significantly smaller residual densities, more precise bond lengths, and last but not least, easier identification of missing or incorrectly assigned atoms.

[1] B. Dittrich, C. B. Hübschle, K. Pröpper, F. Dietrich, T. Stolper, J. J. Holstein, *Acta Crystallogr. Sect. B* 2013, *69*, 91.
[2] J. Lübben, C. Wandtke, C. B. Hübschle, M. Ruf, B. Dietrich, *Acta Crystallogr. Sect. A,* 2019, 75.

Figure 1

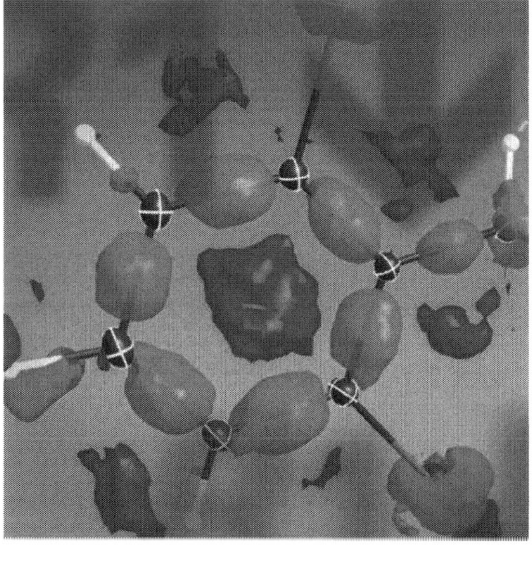

S17-6
Aspherical scattering factors from multipolar model for X-ray and electron crystallography

P. M. Dominiak[1], M. L. Chodkiewicz[1], B. Gruza[1], K. K. Jha[1], M. Kulik[1], P. Kumar[1], P. Rybicka[1]
[1]University of Warsaw, Biological and Chemical Research Centre, Department of Chemistry, Warsaw, Poland

Atomic electron densities and electrostatic potentials are not spherical and associated with them net charges are rarely close to the formal ones in crystals. Independent atom model (IAM) was very successful in X-ray crystallography over the past 100 years. Nowadays, with better equipment and higher computation power, more accurate models can be used.

Transferable Aspherical Atom Model (TAAM) build from atoms in multipolar formalism [1] and parametrized with a pseudoatom databank (ELMAM [2], Invariom [3] or UBDB [4]) is one of the alternative models to IAM.To ease the usage of TAAM in crystal structure refinement we interfaced our new software library DiSCaMB [5] with Olex2 [6] program.

Results of TAAM refinements on large X-ray data set clearly show [7] all advantages and disadvantages of TAAM comparing to IAM and the other alternative approach - Hirshfeld Atom Refinement HAR [8]. There are also very encouraging results of TAAM modeling and refinements with electron diffraction data [9].

TAAM becomes a reliable, fast and easy-to-use method for obtaining crystal structures of high accuracy and precision.

Support of this work by the National Centre of Science (Poland) through grant OPUS No.UMO-2017/27/B/ST4/02721 is gratefully acknowledged.

[1] Hansen & Coppens Acta Cryst. A 1978. 34, 909–921; Stewart, Acta Cryst. A 1976, 32, 565–574.
[2] Domagała, et al. Acta Cryst. A 2012, 68, 337 351.
[3] Dittrich, et al. Acta Cryst. B 2013, 69, 91–104.
[4] Kumar, et al. Acta Cryst. A 2019, 75, 398-408.
[5] Chodkiewicz, et al. J. Appl. Cryst. 2018, 51, 193 199.
[6] Dolomanov, et al. J. Appl. Cryst. 2009, 42, 339-341.
[7] Jha, et al. Acta Cryst. B 2019, submitted.
[8] Woińska, et al. Sci. Adv. 2016; 2, e160019.
[9] Gruza, et al., Acta Cryst. A 2019, under review.

Lightning talks of young crystallographers

LT-1

Dicarboxylic Acids: Versatile ligands for metal-organic synthesis.

M. Kremer[1], U. Englert[1]
[1]RWTH Aachen University, Institute of Inorganic Chemistry, Aachen, Germany

How does crystal field stabilization in [d8]-Ni(II) and its absence in [d10]-Zn(II) change the coordination of 2,6-dipicolinic acid and its derivatives, and is there a correlation between the formation of coordination polymers and temperature dependent phase transition of these compounds? How does the coordination behaviour of the different stereoisomers of tartaric acid change in the presence of additional chelating ligands and how does the presence of these ligands change the physical properties and crystallization behaviour of similar carboxylates like aspartates?

These are just some of the questions we will answer in this presentation. We present the results of our current research in the field of BioMOFs, more specifically into the design and synthesis of metal-organic frameworks, or coordination polymers based on non-toxic starting materials and solvents. We chose dicarboxylic acids as ligands for their great versatility, as well as their decent coordination behaviour.

First results have already been published, [1] but we will also present crystal structures never reported previously. Amongst these new compounds are coordination polymers as well as discrete molecules.

In addition to extending the library of known metal-organic compounds featuring dicarboxylic acids, we were able to make several interesting observations connected to the fundamental coordination behaviour of our ligands.

In our presentation, we attempt to provide a systematic overview of our work so far, as well as a detailed look into some of the more interesting recent results.

[1] M. Kremer, U. Englert: Zn and Ni complexes of pyridine-2,6-dicarboxylates: crystal field stabilization matters!, *Acta Cryst.*, 2019, *E75*, 903-911.

Figure 1

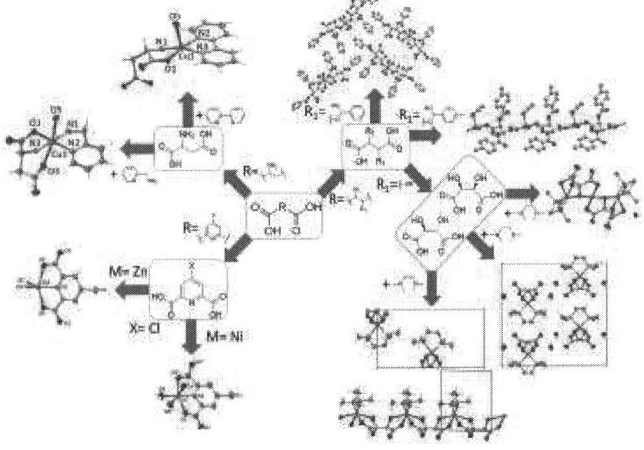

LT-2

Ferromagnetic Interaction in a New Semi-rigid Tricarboxylate-Bridged Ni^{2+} Complex

Y. An[1,2], L. Lu[1], E. Ullrich[2], M. Zhu[1]
[1]Shanxi University, Institute of Molecular Science, Taiyuan, China
[2]RWTH Aachen University, Institute of Inorganic Chemistry, Aachen, Germany

Coordination polymers (CPs), as an extensive class of inorganic–organic hybrid crystalline materials, have attracted considerable attention to be extensively studied for their intriguing structural varieties and potential applications. Over the past decade, the design and synthesis of CPs with desired magnetic properties has become extremely important. A large number of CPs with ferromagnetic, antiferromagnetic, spin canting, metamagnetic, and single chain magnet behavior have been reported.[1-3]

To evaluate the magnetism of complexes containing magnetic metal-organic chains, the new Ni^{2+} complex $[Ni(\mu\text{-}HL^{2-})(2,2'\text{-}bpy)(H_2O)]_n$ (**1**) (H_3L=5-(4-carboxybenzyloxy)isophthalic acid and 2,2'-bpy=2,2'-bipyridyl), was synthesized by hydrothermal method with the help of N-donor auxiliary ligand. Complex **1** shows a trans-conformation wave-like 1D chain with a Ni···Ni distance of 10.314(6) Å, in which the neighboring HL^{2-} ligands adopt alternating positions. The magnetostructural analysis discloses ferromagnetic behavior for the metal cations between 2-300 K. In complex **1**, the *syn–anti* carboxyl group of HL^{2-} and metal-organic chains bridged by HL^{2-} can transmit magnetic interactions. The magnetic data can be well fitted to the Heisenberg Hamiltonian equation: $\hat{H} = -J\hat{S}_1\hat{S}_2$[4], considering molecular field approximation and zero-field splitting effects (ZFS).

FIGURE

Figure 1. Synthesis and ferromagnetism of complex **1**.

ACKNOWLEDGEMENTS
The authors thank the financial support by China Scholarship Council, Shanxi University (scholarship for Yanyan An) and the Natural Science Foundation of China (Grant No. 21571118 & 21671124).

REFERENCES

[1] S. Goswami, G. Leitus, B. K. Tripuramallu, I. Goldberg, *Crystal Growth & Design*. 2017, *17*, 4393-4404.
[2] X. H. Zhang, Z. M. Hao, X.M. Zhang, *Chem. Eur. J.* 2011, *17*, 5588-5594.
[3] J. Boeckmann, C. Nather, *Dalton Trans.* 2010, *39*, 11019-11026.
[4] Y. Liu, N. Li, L. Li, H. L. Guo, X. F. Wang, Z.X. Li, *CrystEngComm*. 2012, *14*, 2080-2086.

Figure 1

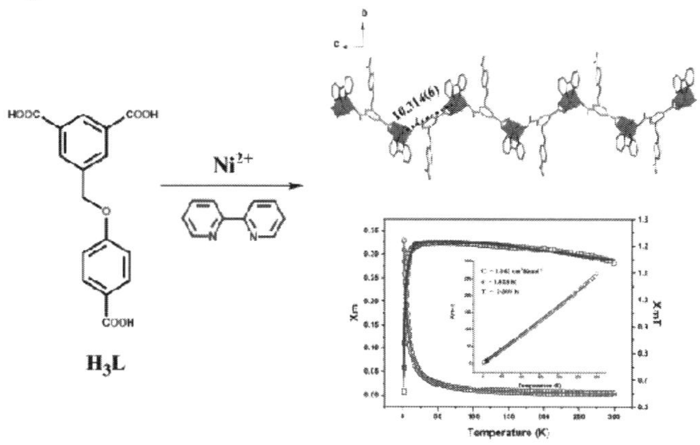

[Ni(μ_2-HL^{2-})(2,2'-bpy)(H$_2$O)]$_n$
1D
ferromagnetism

LT-3
Dihydroorotases from pathogenic bacteria

J. Sławek[1], D. Miks[2], I. G. Shabalin[2], W. Minor[2], K. Lewiński[1]
[1]Jagiellonian University, Faculty of Chemistry, Kraków, Poland
[2]University of Virginia, Charlottesville, Virginia, United States

In recent years antimicrobial resistance becomes one of the biggest problems faced by researchers all over the world. As a response to environmental changes, microorganisms may quickly adopt and spread the resistance genes. Because of that, many antibiotics become less and less effective and new therapies are urgently needed.

The aim of this study was to obtain and characterize the crystal structure of two dihydroorotases from pathogenic bacteria *Yersinia pestis* and *Vibrio cholerae*. Dihydroorotase (DHO) is the third enzyme of *de novo* pyrimidine biosynthesis pathway which catalyzes the interconversion of N-carbamoyl-l-aspartate to 4,5-dihydroorotate. The substantial difference between bacterial and mammalian DHOs makes it a promising drug target for disrupting bacterial growth.

Recombinant DHOs were overexpressed in *E. coli* and purified using affinity chromatography. Proteins were crystallized using the sitting drop technique. Crystallization experiments were set up using commercially available screen solutions and all crystallization drops were equilibrated against a 1.5 M NaCl in the reservoirs. Crystallization conditions were further optimized using several approaches.

In determining the two structures the key factor was to implement the small molecules as additives to the crystallization drops. Both enzymes are homodimers build of chains containing a $(\beta/\alpha)_8$-barrel connected to an adjacent domain. In both structures the active site is highly conserved, including the positions of all catalytically important residues and modified lysine residue.

The analysis revealed that both DHOs are close homologs with *E. coli* DHO. Based on the sequence and structure similarities, both proteins were classified as members of bacterial type II DHOs, which is genetically distinct from the human DHO subunit. The differences in bacterial and human DHOs may be used for designing effective therapies that eliminate pathogens with the minimum adverse effect of the host.

LT-4
NASICON-$Na_3Fe_{2-y}Mn_y(PO_4)_3$ cathode materials for Na-ion batteries – comprehensive analysis of the relationship between structural and electrochemical properties using HT-XRD and operando-XRD studies

K. Walczak[1], B. Gędziorowski[1], A. Kulka[1], J. Molenda[1]
[1]AGH University of Science and Technology, Faculty of Energy and Fuels, Department of Hydrogen Energy, Kraków, Poland

In the last two decades the fast growing demand for Li-ion batteries (LiBs) powering the electronic devices, cars and smart grids is observed, what is a reason for declining global reserves of Li, and increasing its price. Sodium-ion batteries seem to be possible alternative for LiBs, especially in large-scale energy storage and the safe cathode and anode materials for them are being quested.

The aim of this work is deep analysis of structural and electrochemical properties of the NASICON-$Na_3Fe_{2-y}Mn_y(PO_4)_3$ (y=0, 0.1, 0.2, 0.3, 0.4) group - cathode materials for Na-ion batteries.

The samples were investigated using HT-XRD measurements in temperature range of 25°C–300°C. To study the electrochemical properties of the Na|Na$^+$|$Na_3Fe_{2-y}Mn_y(PO_4)_3$ cells, the operando-XRD measurements were conducted for the samples with y=0 and y=0.2.

NASICON-$Na_3Fe_{2-y}Mn_y(PO_4)_3$ crystallize in *C2/c* symmetry and undergo two phase transitions at ~95°C and 145°C, resulting in reduced monoclinic distortion. For y=0.2 and y=0.3 samples, the distortion is smallest, which perfectly correlates with the electrochemical properties of Na|Na$^+$|$Na_3Fe_{2-y}Mn_y(PO_4)_3$ cells, indicating the best kinetics. The conducted in-situ XRD measurements showed two-phase mechanism of sodium insertion/deinsertion, followed by the reaction:

$$Na_3Fe_2(PO_4)_3 - xNa^+ + xe^- \leftrightarrow (1-x)Na_3Fe_2(PO_4)_3 + xNa_4Fe_2(PO_4)_3$$

Presented analysis of structural properties of NASICON-$Na_3Fe_{2-y}Mn_y(PO_4)_3$ cathode materials is an important factor in understanding the electrochemical processes during the cell operation. This work shows that the modification of the chemical composition of the cathode materials has a major impact on electrochemical properties and allows to design eco-friendly, inexpensive NASICON-type cathodes for Na-ion batteries.

This work was funded by the Polish Ministry of Science and Higher Education (MNiSW) based on decision number 0020/DIA/2016/45.

K.Walczak et al. ACS, Appl. Mat. And Interfaces, doi.org/10.1021/acsami.9b10184

LT-5

New zinc coordination compounds as effective luminophores and precursors of ZnO nanoparticles

M. Świątkowski[1], R. Kruszyński[1]
[1]Łódź University of Technology, Institute of General and Ecological Chemistry, Łódź, Poland

Nowadays, a designing of coordination compounds exhibiting strictly defined features is one of the most important fields of material engineering [1, 2]. Structural investigations allow linkage between the structure of the compound and its physicochemical properties [3]. Such relationships are the basis for the intentional application of coordination compounds in practice.

The aim of this work was the design and synthesis of new zinc coordination compounds, which can serve as effective luminophores as well as precursors in the synthesis of ZnO nanoparticles.

Coordination compounds were synthesized via direct reactions between zinc carboxylates (butyrate, isobutyrate, and valerate) and 2,2"-bipyridine (bpy). These syntheses led to the formation of three new coordination compounds of general formula [Zn(carboxylate)(bpy)(H2O)](carboxylate). They were structurally characterized by X-ray crystallography. The synthesized compounds exhibit strong fluorescence, which originates from coordination entities and its main source is bpy. Carboxylate anions coordinating to zinc cations differentiate emission properties of compounds in terms of location of global emission maximum and its intensity.

Zinc oxide nanoparticles were produced from coordination compounds via the single precursor method utilized in two experiments. SEM study revealed that morphology and size of ZnO nanoparticles are strictly related to the conversion conditions and precursor structures.

Summarizing, a government of the structure of coordination compounds by the introduction of small alterations (e.g. change of carboxylate anions) is a tool for the intentional modification of their emission properties and habit of nanoparticles produced by their use.

References

[1] T. Rasheed, F. Nabeel, *Coordination Chemistry Reviews,* 401 (2019) 213065.
[2] K. Liu, W. Shi, P. Cheng, *Dalton Transactions*, 40 (2011) 8475.
[3] M .Swiatkowski, R. Kruszynski, *Applied Organometallic Chemistry,* 33 (2019) 4812.

LT-6

Towards better atomic displacement parameters in structural macromolecular models from micro-electron diffraction

M. Kulik[1], M. L. Chodkiewicz[1], P. M. Dominiak[1]
[1]University of Warsaw, Biological and Chemical Research Centre, Department of Chemistry, Warsaw, Poland

Electrostatic Coulomb potential maps are obtained with methods using electron beams for structure determination, such as micro-electron diffraction and cryo-electron microscopy. These maps, at atomic or near-atomic resolution, are affected by multiple factors, including atomic partial charges or thermal motion of the atoms in the sample. However, the effect of those factors on the potential maps is not well studied. It is expected that the atomic displacement parameters should be bigger than presently observed in refined electron diffraction structural models [1]. Here, we investigate the properties of electrostatic potential maps calculated using aspherical atom databank.

Databanks of aspherical atom types, such as University at Buffalo Data Bank (UBDB) [2], have been created since electron density of atom types are transferable between different molecules in similar chemical environment. Next, the pseudoatom types can be used to recreate the electron density distribution of macromolecules via calculating structure factors [3] and to calculate the accurate electrostatic potential maps. UBDB reproduces the molecular electrostatic potential of molecules within their entire volume better than the simple point charge models used in molecular mechanics or neutral spherical models used in electron crystallography. In this study, micro-electron diffraction data of a lysozyme system is used for calculation of the electrostatic potential maps with increased or diminished atomic displacement parameters. This allows us to compare their distribution with experimental micro-electron diffraction maps and facilitates the interpretation of the physical meaning of atomic displacement parameters with respect to their well-established meaning in X-ray crystallography.

The authors acknowledge NCN UMO-2017/27/B/ST4/02721 grant.

References

[1] Gruza et al. Acta Cryst. A, under review
[2] Kumar et al. (2019). Acta Cryst. A75, 398-408
[3] Chodkiewicz et al. (2018). J. Appl. Cryst. 51, 193-199

LT-7

Polymorphism and resulting luminescence properties of 1-acetylpyrene

D. Tchoń[1], D. Trzybiński[2], A. Wrona-Piotrowicz[3], A. Makal[1]
[1]University of Warsaw, Department of Chemistry, Warsaw, Poland
[2]University of Warsaw, Biological and Chemical Research Centre, Faculty of Chemistry, Warsaw, Poland
[3]Łódź University of Technology, Department of Chemistry, Łódź, Poland

Policyclic Aromatic Hydrocarbons have recently been attracting attention due to their potential applications as optoelectronic materials[1] or luminescent markers.[2] The relative ease of chemical modification, sensitivity to environment in solution[3] and ability to form luminescent aggregates in solid state[4] makes them a potent target for applications-oriented studies.

While a lot of attention is paid towards seeking new efficient and stable organic luminophores, model compounds such as acetylpyrene (1AP) appear to be neglected. Until very recently only one crystal structure of 1AP has been known.[5] Our study aims to correlate crystal packing and luminescence properties observed in various polymorphs of 1AP and other poliacetylpyrenes.

New crystals of 1AP have been found by the means of recrystallization from several solvent mixtures. Samples were investigated by the means of UV-Vis excitation and emission spectroscopy as well as single crystal X-ray diffraction. Resulting geometries and spectra were compared with results of DFT calculations. High-pressure experiments for selected specimens are ongoing.

Experiments performed to date revealed two new crystal forms of 1AP. New-found samples feature different packing scheme and luminescence, the yield and spectral range of which differ from that of known polymorph. Obtained data shows that spectroscopic properties of poliacetylpyrenes in solid state can be modified not only by the means of gradual acetylation but also by control over crystallization conditions.

This study was financially supported by the National Science Centre Poland (NCN) based on decision DEC-2015/17/B/ST4/04216.

[1] T. M. Figueira-Duarte and K. Müllen, *Chem. Rev.* 2011, 111, 7260-7314
[2] I. O. Aparin et al., *J. Org. Chem.* 2017, 82, 10015-10024
[3] Y. Niko et al., *Tetrahedron* 2012, 68, 6177-6185
[4] A. Hayer et al., *J. Phys Chem. B* 2009, 110, 7653-7659
[5] S. K. Rajagopal et al., *Chem. Commun.* 2014, 50, 8644-8647

Figure 1

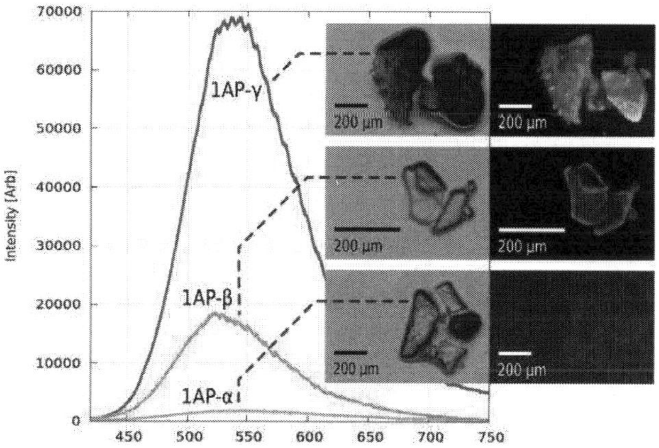

LT-8

Understanding cation distribution in $Zn_{1+x}GeN_2O_x$ (x < 0.1)

Z. Wang[1,2], J. Breternitz[1], A. Franz[1], S. Schorr[3,2]
[1]Helmholtz-Zentrum Berlin für Materialien und Energie GmbH, Structure and Dynamics of Energy Materials, Berlin, Germany
[2]Freie Universität Berlin, Institute of Geological Sciences, berlin, Germany
[3]Helmholtz-Zentrum Berlin für Materialien und Energie GmbH, Structure and Dynamics of Energy Materials, Berlin, Germany

Increasing interest in photovoltaic devices has motivated researchers to develop new absorber materials for solar cells. Zn-IV-N_2 (IV: Si, Ge, Sn) materials are being considered as potential candidates, since they are earth-abundant and non-toxic alternatives to group-III nitride materials.

Zn-IV-N_2 presents as a derivative of group-III nitrides, where the trivalent cations are alternatingly replaced by divalent (Zn^{2+}) and tetravalent cations. From a structural point of view, the additional cation type allows for two possible crystal structures: the wurtzite aristotype, which necessarily exhibits cation disorder as it only contains one cation site, or the β-$NaFeO_2$-type structure, which allows cation ordering. It has been postulated through DFT calculations that Zn-IV-N_2 materials show a distinct bandgap tuning mechanism through cation disorder in addition to the possibility of cation alloying.[1,2] Besides, it has been shown that oxygen incorporation into $ZnGeN_2$ can trigger cation disorder, since oxygen-rich phases (x > 0.1) crystallise in the wurtzite-type structure.[3]

The main purpose of this work is to deconvolute the contributions of oxygen to cation disorder and bandgap tuning from effects independent of oxygen content. We synthesised $Zn_{1+x}GeN_2O_x$ powder with a variable oxygen content to cover the spectrum of ordered and disordered materials. Samples have been characterized by chemical analysis and X-ray diffraction (XRD). Neutron diffraction was used to characterise the cation distribution in this material, since Zn^{2+} and Ge^{4+} are isoelectronic and hence indistinguishable by XRD. Furthermore, we employed UV-VIS spectroscopy to determine the materials bandgap energy. Oxygen-poor $Zn_{1+x}GeN_2O_x$ crystallizes in the β-$NaFeO_2$-type structure showing about 10 % Zn-Ge disorder on the two structural cation sites. This cation disorder engineering through the oxygen incorporation achieves a bandgap tuning between 2.7-3.5 eV.

References

[1] P. Narang, et al., *Adv. Mater. 26*, 2014, 1235
[2] E.W. Blanton, et al. *Cryst. Growth 461*, 2017,38
[3] J. Breternitz, et al., *Phys. Status Solidi (a)*, 2019, 1800885

LT-9

Structures of Phosphorylated Hydroquinolinols: Molecular and Supramolecular Aspects

A. Pietrzak[1], J. Koszuk[1], T. Janecki[1], W. Wolf[1]
[1]Łódź University of Technology, Łódź, Poland

Phosphonates are important class of compounds for medicine, chemical industry and agriculture. Especially, they are also invaluable reagents for various synthetic routes, based on Horner-Wadsworth-Emmons (HWE) methodology. Presented phosphorylated hydroquinolinols contain phosphoryl, hydroxyl and toluenesulfonyl (Ts) functional groups linked to unsaturated nitrogen containing cyclic moiety. It makes these compounds prone to form variety of intra- and intermolecular hydrogen bonds.

The major objective of the study is to characterize resonance assisted hydrogen bond (RAHB) between phosphoryl and hydroxyl groups and determine its influence on supramolecular assembly. RAHBs were reported in several C,H,N,O compounds. However, number of reports on these bonds in organic phosphorus compounds (RAHB-P) is very scarce.

Single crystals suitable for Xray diffraction studies were recrystallized by slow evaporation from various solvents. X-ray diffraction experiments were collected on a XtaLAB Synergy, Dualflex, Pilatus 300K diffractometer at low temperature (100 K). Structural data is supported by theoretical calculations *i.e.* Hirshfeld Surface with mapped electrostatic potential, Fingerprint Plots analysis and intermolecular interaction energy calculations done with the methodology implemented in the *CrystalExplorer17* program.

Investigated compounds form intramolecular hydrogen bond defined by S(6) graph set notation. Geometrical parameters and computation results defining this intramolecular effect suggest that the molecular conformation is constrained by strong RAHB-P. Additionally, the conformation is stabilized by intramolecular hydrogen bonds provided by Ts group being a driving force for the supramolecular assembly. Strong intramolecular RAHB-Ps observed in studied crystals play crucial conformation-controlling role. This feature together with the dual role of Ts group makes studied compounds potential building blocks for crystal engineering strategies.

This work is supported by National Science Centre, Poland (NCN) according to decision 2019/03/X/ST4/01389.

LT-10
In situ crystallization and structural investigation of binary cocrystals of diamines and diols

G. Cichowicz[1], M. Cyrański[1], R. Boese[1], Ů Dobrzycki[1]
[1]University of Warsaw, Warsaw, Poland

Simple amines and alcohols are usually used as "building blocks" in supramolecular synthesis, solvents, ligands or cocrystallization agents. Variety of applications originates in presence of functional groups able to participate in strong hydrogen bonds. As both hydroxyl and amine moieties are complementary, designing new cocrystals of such compounds can lead to complex structures with large diversity of crystal architectures.

Cocrystals of primary monoalcohols and amines exhibits a topology similar to allotropic form of gray arsenic - layers build by six-membered ring-shaped with chair conformation [1]. In order to obtain complex three-dimensional motifs, cocrystallizations of linear aliphatic diamines and diols with different chain length were performed. Most of them are liquids at ambient conditions, therefore an IR laser supported *in situ* crystallization method has been applied [2]. In most of obtained phases containing saturated diols characteristic layers composed of six-membered, ring-shaped hydrogen bond motif is present. However, in some phases, like in the case of 1,5-diaminopentane with 1,5-pentanediol co-crystal, this motif does not occur, what is associated with a conformational change of both molecules. These structures contain molecules arranged in 1D columns instead. Relatively small change in shape of a molecule may affect the final structure. Indeed, cocrystallization of unsaturated 1,4-but-2-endiol with 1,6-diaminohexane resulted in formation of incommensurately modulated phase. Among all obtained crystal structures, six-membered ring-shaped arrangements exhibits highest values of melting point temperatures in comparison to other motifs.

The National Science Center in Poland (Grant SONATA BIS 6 NCN, 2016/22/E/ST4/00461) is gratefully acknowledged for financial support.

[1] O. Ermer, A. Eling, *J. Chem. Soc. Perkin. Trans.*, 1994, 2, 925-444.
[2] R. Boese, *Z. Kristallogr.*, 2014, 229, 595-601.

LT-11
Allogon isomerism in a series of high spin Fe(II) complexes

B. Braun Cula[1]
[1]Humboldt Universität zu Berlin, Institute of Chemistry, Berlin, Germany

Allogon isomerism for molecular complexes is a rare case of geometric isomerism.

Fe^{II} centres in square planar coordination spheres and high-spin configuration is a rarely observed situation in coordination chemistry.

We were fortunate to come across a class o Fe complexes which show allogon isomerism. Blue crystals reveal the tetrahedral geometry whereas the pink crystals the perfect square planar geometry.

Core motiv (tetrahedral, blue crystals- Picture 1)

Core motiv (square planar, pink crystals- Picture 2)

Figure 1

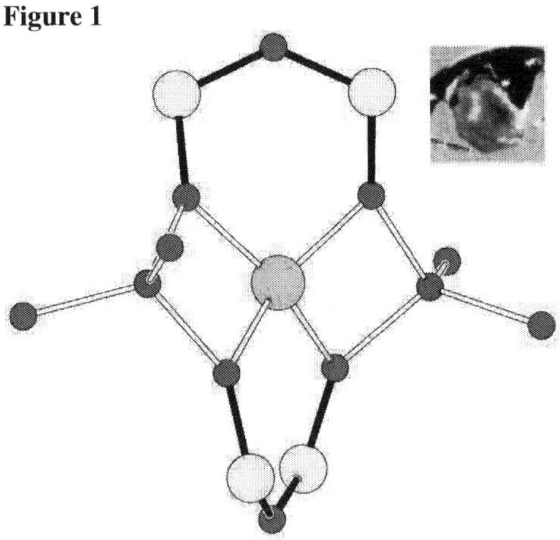

Figure 2

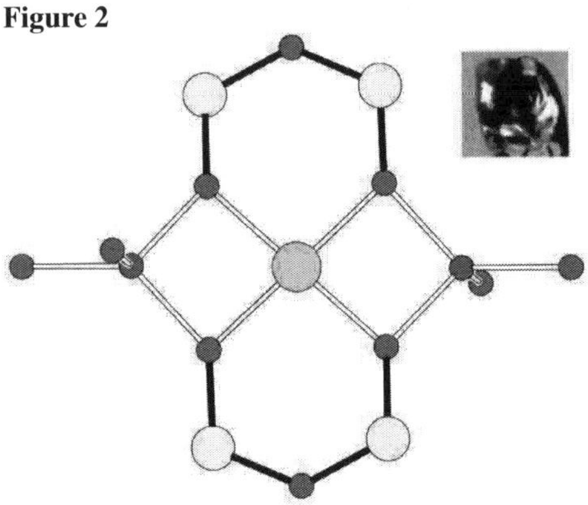

LT-12
Structural and Macroscopic Investigation of CrAs at Low Temperatures and High Pressures

A. Eich[1], A. Grzechnik[2], T. Müller[3], C. Paulmann[4], K. Friese[1]
[1]Research Center Jülich GmbH, Jülich Centre for Neutron Science-2 (JCNS-2), Jülich, Germany
[2]RWTH Aachen University, Institute of Crystallography, Aachen, Germany
[3]Forschungszentrum Jülich GmbH, Jülich Centre for Neutron Science, Garching, Germany
[4]University of Hamburg, Mineralogisch-Petrographisches Institut, Hamburg, Germany

Chromium Arsenide (CrAs) is the first Cr-based superconductor, exhibiting pressure-induced superconductivity with a maximum $T_c = 2.2$ K at around 1 GPa [**1**,**2**] and the dome-like shaped superconducting phase region existing in the vicinity of a magnetically ordered state. The magnetic structure is described as double helices propagating along the *c** direction with the spins in the (*a*,*b*) plane. The first-order phase transition from the helimagnetically ordered antiferromagnetic state to a paramagnetic state is clearly and consistently observed in our macroscopic measurements of magnetization, resistivity and heat capacity at $T_N \approx 267$ K. The transition is furthermore connected to an abrupt increase in unit cell volume of about 2.4%, though the symmetry of the crystal structure (MnP-type *Pnma*, $Z = 4$) does presumably not change [**3**].

As the understanding of the behaviour of the crystal structure of CrAs so far is based primarily on powder data, the objective of our work is to precisely determine the crystal structure by single crystal diffraction at low temperatures as well as at high pressures. This knowledge will serve as base for further studies of the crystal and magnetic structures of CrAs at simultaneously low temperatures and high pressures, in particular within or in the vicinity of the superconducting phase.

For this, synchrotron x-ray diffraction experiments were performed on single crystals between 300 K and 20 K at ambient pressure, and between 0.92 GPa and 9.45 GPa at room temperature.

The preliminary refinements show a good agreement with the literature data for the room temperature phase. Below the transition temperature the symmetry *Pnma* is preserved and the reflections can still be indexed with the orthorhombic cell, although a deterioration of the crystal quality is observed.

[**1**] R.Y. Chen, N.L. Wang, *Rep. Prog. Phys.* **82**, 012503 (2019)
[**2**] W. Wu, *et al.*, *Nat. Commun.* **5**, 5508 (2014)
[**3**] T. Suzuki, H. Ido, *J. Appl. Phys.* **69**, 4624 (1991)

LT-13

DHS structure and function. On the crossroads between polyamines and posttranslational modification.

E. Wątor[1], P. Wilk[1], P. Grudnik[1]
[1]Jagiellonian University, Małopolska Centre of Biotechnology, Kraków, Poland

Deoxyhypusine synthase (DHS) is a transferase catalysing the formation of deoxyhypusine, which is the first, rate-limiting step of unique post-translational modification: hypusination. DHS catalyzes the transfer of 4-aminobutyl moiety of spermidine to a specific lysine of eIF5A precursor in a NAD-dependent manner. This modification occurs exclusively on only one protein: eukaryotic initiation factor 5A (eIF5A) and it is essential for cell proliferation. Malfunctions of hypusination pathway, including those caused by mutations within DHS encoding gene, are associated with such conditions as cancer or neurodegeneration. The main goal of the presented study was to structurally uncover the first step of hypusination and its pathological abnormalities. DHS wild type and its two mutants were expressed, purified and crystallization attempts were undertaken. Crystallization trials have led to the determination of five high-resolution crystal structures of DHS wt in apo form and in complexes with natural substrates. Additionally, two crystals structures of N173S DHS were determined allowing for detailed analysis and comparison of the wild-type and physiologically relevant mutant. Determined high-resolution crystal structures show a detailed binding mode of natural ligands to DHS. Additionally, complementary analysis of protein oligomerization, activity and stability in solution further support the conclusions drawn from crystallographic studies.

Information drawn from these structural studies is a significant step forward in understanding the molecular basis of the first step of hypusination and its rare, but severe in consequences malfunctions.

Figure 1. A. DHS WT protein crystals in the crystalization drop visiualised under polarised light. B. Crystal on the micro-mount in the X-ray beam. The green scale bars shows approximate size of the sample. C. Diffraction pattern of DHS crystal recorded at BESSY-MX BL 14.1.

Figure 1

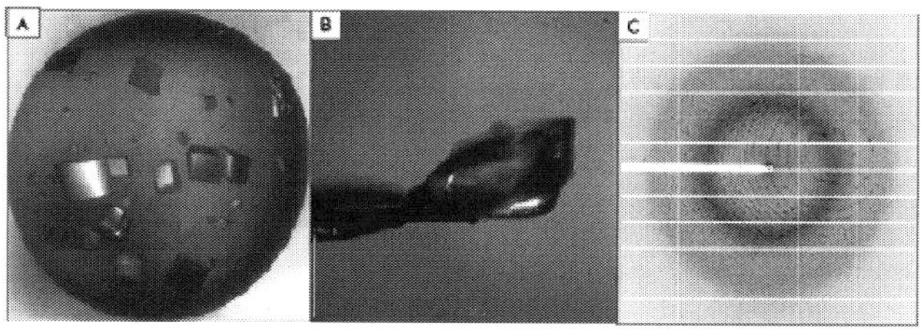

LT-14

Preferential orientation of $Ce_{1-x}Ln_xO_{2-y}$ nano-sized crystallites in star-shaped hierarchical porous particles.

P. Woźniak[1,2], M. A. Małecka[1,2]
[1]Institute of Low Temperature and Structure Research, Polish Academy of Sciences, Department of Nanomaterials Chemistry and Catalysis, Wrocław, Poland
[2]Institute of Low Temperature and Structure Research, Polish Academy of Sciences, Wrocław, Poland

Cerium oxide in nanocrystalline form possess unique properties that differs from their bulk material counterparts. These are the concentration of edges and corners or the ratio of surface to volume [1]. Introduction of dopants into the material structure modifies oxygen vacancy defect concentration, thus improving its redox properties [2]. In addition, engineering porosity of the material increases its surface area enabling catalytic reactions to occur more easily [3]. Joining together this three characteristics by creating hierarchical porous material was the aim of the present study.

The particular objective was to synthesize hierarchically structured micron-sized particles with expectable increased degree of porosity, composed of $Ce_{1-x}Ln_xO_{2-y}$ (Ln=lanthanides) mixed oxide nanocrystallites for prospective applications as active catalytic media.

Solvothermal method was employed to obtain $Ce_{1-x}Ln_x(HCOO)_3$ (x=0.1) crystal agglomerates that were subsequently subjected to oxidative thermolysis that enabled the formation of $Ce_{1-x}Ln_xO_{2-y}$ nano- to micro- hierarchically structured crystalline material. PXRD, SEM, EDX, TEM, HRTEM, SAED and TG has been used to characterize as synthesized samples.

Star-shaped particles with sizes ranging from 5 to 10 μm composed of mixed oxide $Ce_{1-x}Ln_xO_{2-y}$ (x=0.1) nanocrystallites (~10 nm) were formed. Material showed three-level architectural hierarchy (Fig. 1). TEM images and SAED patterns showed preferential orientation of nanocrystallites in particle arms. SEM imaging allowed to gain insight into porous texture of material.

Synthetic protocol developed in this research allows to produce hierarchically structured particles composed of mixed oxide cerium-lanthanide nano-sized crystallites.

Fig. 1. 3-level hierarchical organization of CeO_{2-y} media.

References

[1] M. Fernandez-Garcia et. al., Chem. Rev. **2004**, 104, 4063-4104.
[2] D. Mukherjee et. al., Catal. Today **2018**, 309, 227-235.
[3] W. Schwieger et. al., Chem. Soc. Rev., **2016**, 45, 3353.

Figure 1

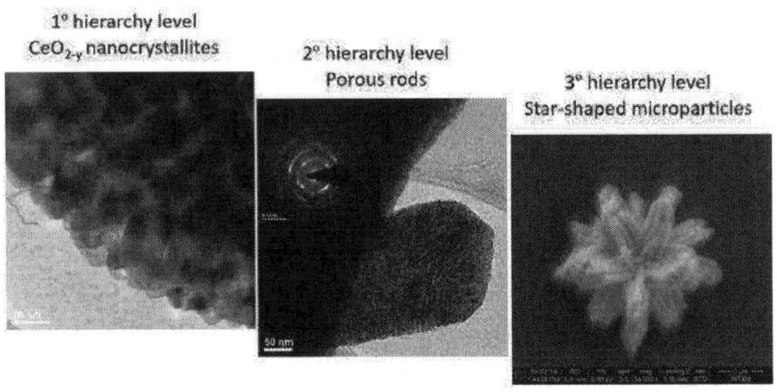

LT-15

A study of transport properties at the single-crystal based charge transfer interfaces

B. Debnath[1], M. Bretschnieder[1], S. J. Wang[1,2], M. Knupfer[1], B. Büchner[1], Y. Krupskaya[1]
[1]Leibniz Institute for Solid State and Materials Research Dresden, Institute for Solid State Research, Dresden, Germany
[2]Dresden Integrated Center for Applied Physics and Photonic Materials (IAP), Physics, Dresden, Germany

With the recent progress in organic electronics, charge transfer interfaces become significant research interest to study and understand the fundamental properties of organic semiconductors. To understand the mechanism of charge transfer, single-crystal based charge transfer interface devices are proved to be a very promising tool[i,j,k]. In recent studies[i,j,k], it has been observed that the electrical properties at the interfaces of two organic semiconductor materials can differ noticeably from those of the individual materials, which are initially insulators. In this work, we investigate single-crystal based charge transfer interfaces where Rubrene, in form of single crystals, is used as a donor material. As an acceptor, we use one of the new striking acceptor materials, F6-TCNNQ, which is deposited on top of Rubrene crystals via thermal evaporation. We observe strongly enhanced electrical conductivity at the interface. Moreover, temperature dependent as well as Hall-effect measurements are performed to investigate the charge transfer effects at their interfaces.

This work is financially supported by DFG (KR 4364/4-1)

[i]I. G. Lezama et al., *Nature Materials* 11, 788–794 (2012)
[j]Y. Krupskaya et al., *Adv. Funct. Mater.* 26, 2334 (2016)
[k]Y. Krupskaya et al., *Adv. Mater. Interfaces* 3, 1500863 (2016)

LT-16

Intermolecular Orbital Interactions in the Structures with Systems Containing Aromatic N-Heterocyclic Rings

T. Sierański[1]
[1]Łódź University of Technology, Institute of General and Ecological Chemistry, Łódź, Poland

Systems with N-heterocyclic aromatic rings may be engaged in a number of different interactions[1] which are thought to arise from many sources including overlapping of molecular orbitals. Although a lot has been learnt so far there is a lack of knowledge on the hierarchy of orbital interactions in the compound structures together with the factors determining this hierarchy and the orbitals engaged in particular interactions.

In order to learn of the factors defining the type of orbital interactions and to know how these interactions may engage in the stability of a particular compound structure the extensive study covering model N-heterocyclic homodimers (pyrrole, purine, pyridine, imidazole, pyrazole, 1,2,3-triazole, 1,2,4-triazole and quinoline) has been performed.

All calculations have been carried out using B3LYP-D3BJ[2] and M06-2X[3] density functionals and were performed on the mentioned homodimers differentiating many geometrical parameters. The calculations covered interaction energies between monomers, natural bond orbital and energy decomposition analyses. The obtained results were associated with the data gathered from Cambridge Structural Database.[4]

Generally, with the increase in the shift of one monomer along the plane of the other the contribution of the $\pi \rightarrow \pi^*$ diminishes and other interactions appear (e.g. $\sigma_{C\text{-}H} \rightarrow ry^*_H$ and $\sigma_{C\text{-}H} \rightarrow \sigma^*_{C\text{-}H}$). The role of $\sigma_{C\text{-}H} \rightarrow \pi^*$ is not as significant as it could be expected. Even for coplanar monomers with geometries for which the formation of N···H hydrogen bonds is not possible, the intermolecular interaction energy is binding and this "in-plane interaction" is an additional binding force stabilizing a structure.

[1] M Guin, GN Patwari, S Karthikeyan, *Phys Chem Chem Phys* 13 2011 5514.
[2] B Brauer, M. K. Kesharwani, S. Kozuch, JML Martin *Phys Chem Chem Phys* 18 2016 20905.
[3] Y Zhao, DG Truhlar, *Acc Chem Res* 41 2008 57.
[4] CR Groom, IJ Bruno, MP Lightfoot. SC Ward, *Acta Cryst* B72 2016 171.

LT-17

CRYSTAL STRUCTURES AND PHASE TRANSITIONS OF IMIDAZOLIUM HYPODIPHOSPHATES

D. Budzikur[1], V. Kinzhybalo[2], K. Ślepokura[1]
[1]Wrocław University of Science and Technology, Faculty of Chemistry, Wrocław, Poland
[2]Institute of Low Temperature and Structure Research, Polish Academy of Sciences, Wrocław, Poland

Hypodiphosphoric acid ($H_4P_2O_6$), first synthesized by Salzer in 1877, has two phosphorus atoms at the +4 oxidation state, which are connected by a direct covalent bond. Studies on $H_4P_2O_6$ and its salts intensified during last few years after the discovery of ferroelectricity in $(NH_4)_2(H_2P_2O_6)$ [1]. Research on imidazolium salts is a continuation of our studies on organic hypodiphosphates [2,3].
Crystals of two imidazolium salts, $(C_3H_5N_2)_2(H_2P_2O_6)$ (**1**) and $(C_3H_5N_2)(H_3P_2O_6)$ (**2**) were obtained in the reactions of imidazole with hypodiphosphoric acid. Compounds were characterized by XRD, PXRD and thermal analysis. DSC revealed two phase transitions in **2**: at about 290 and 310 K. Crystal structures of high- and low-temperature phases (HT, LT) were determined, which showed orthorhombic (*Pnna*, No. 52) → monoclinic ($P2_1/n11$, No. 14) structural change on cooling with a modulated phase in between (the *a*-axis is doubled in LT). The asymmetric unit increases fourfold ($Z' = ½ \rightarrow 2$). Hydrogen-bonded anions form channels in which imidazolium cations are located (resembling honeycomb). The structural analysis of **2** suggests that imidazolium cations change their orientation and the hydrogen atoms in $(H_3P_2O_6)^-$ anions become partially ordered in LT.
Compound **1** crystallizes in tetragonal, chiral space group $P4_32_12$ (No. 96) with one imidazolium cation and a half of $H_2P_2O_6^{2-}$ anion (lying on a twofold axis) in the asymmetric unit. Compared to **2**, the crystal structure of **1** is heavily deformed. One of the

main differences is the parallel arrangement of the imidazolium cations in relation to $H_2P_2O_6^{2-}$ anions (Fig. 1).

Fig. 1. The comparison of crystal packing of **1** and **2** (HT phase).

Literature

[1] P. Szklarz, M. Chański, K. Ślepokura, T. Lis, *Chem. Mater.*, **23** (2011) 1082.
[2] M. Emami, K. A. Ślepokura, M. Trzebiatowska, N. Noshiranzadeh, V. Kinzhybalo, *CrystEngComm*, **20** (2018) 5209.
[3] M. Otręba, D. Budzikur, Ł. Górecki, K. Ślepokura, *Acta Cryst. C*, **74** (2018) 571.

Figure 1

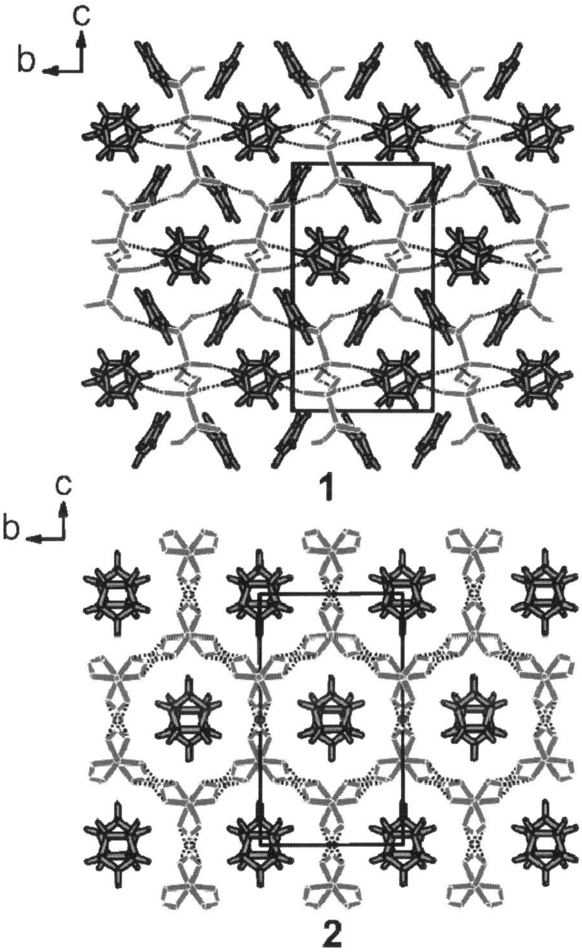

LT-18
Benzonitrile Substituted 1,3-Diketones as Linkers for Heterobimetallic MOFs

U. Englert[1], S. van Terwingen[1]
[1]RWTH Aachen University, Institute of Inorganic Chemistry, Aachen, Germany

Introduction

Metal-organic frameworks (MOFs) have been investigated intensively in the last few decades. Their applications cover a wide variety of fields, such as separation processes, gas storage or catalysis.[1] While they usually incorporate one kind of metal ions, we focus on using ditopic ligands in order to create heterobimetallic MOFs.[2]

Materials and methods

Our newest candidate is 1-(4-cyanophenyl)butan-1,3-dion (HCPB), which has already successfully been used for the O coordination of neodymium and terbium.[3] In contrast to many of our other ligands, it does not exhibit 2*mm* but only *m* local symmetry. This leads to interesting isomerism of its coordination compounds and, thus, their structures in the solid state.

Results and Conclusion

We herein present a new synthesis route for the coordination of 1,3-diketones to rare earth elements yttrium and ytterbium. The use of 1,10-phenanthroline as a co-ligand was found to be an effective way to obtain a defined coordination compound as a crystalline solid. As expected, the rare earth ions adopt a square antiprismatic coordination sphere. Moreover, the iron(III) complex crystallizes predominantly as the *fac*-isomer in which all nitrile moieties point in the same direction as well.

Figure 1: PLUTON plots of $[Y(CPB)_3(phen)]$ (left) and $[Fe(CPB)_3]$ (right).

References

[1] D. Farrusseng, *Metal-Organic Frameworks: Applications from Catalysis to Gas Storage*, Wiley-VCH, Weinheim, 2011.
[2] M. Kremer, U. Englert, *Z. Kristallogr.* 2018, *233*, 437-452.
[3] P. C. Andrews, W. J. Gee, P. C. Junk, H. Krautscheid, J. G. MacLellana, *Chem. Commun.* 2010, *46*, 5948-5950.

Figure 1

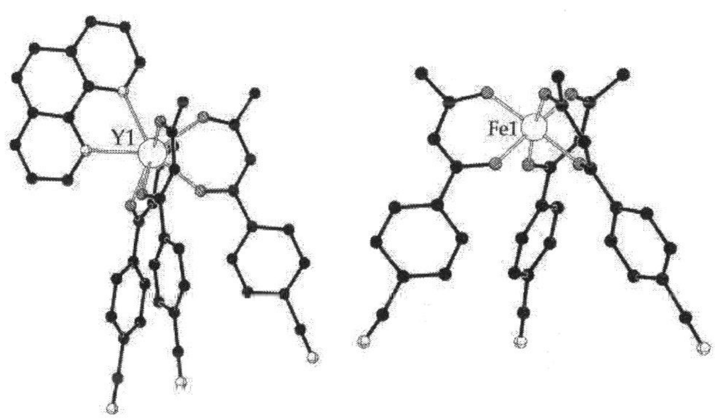

LT-19

Non-linear optical properties and topology of hydrogen bonding patterns of (*S*)-2-amino-3-guanidinopropanoic acid monochloride

P. Rejnhardt[1], M. Daszkiewicz[1], J. K. Zaręba[2]
[1]Institute of Low Temperature and Structure Research, Polish Academy of Sciences, Division of Structure Research, Wrocław, Poland
[2]Wrocław University of Science and Technology, Faculty of Chemistry, Wrocław, Poland

Second harmonic generation (SHG) in organic crystals is a subject of extensive investigation for years. The absence of an inversion centre in crystal is mandatory in order to obtain SHG response, although many other features play an important role and must be taken into account for synthesis of non-linear materials. For example, an occurrence of delocalized π electrons and intramolecular donor-acceptor charge transfer between two molecular subparts (functional groups) is necessary. It leads to large hyperpolarizability β and this kind of materials are good candidates for second harmonic generation.

Here we present crystal and molecular structure of new compound, monochloride salt of (*S*)-2-amino-3-guanidinopropanoic acid, which is an analogue of one of the most important amino acid L-arginine. Crystal structure was determined by X-ray diffraction at room and low-temperature conditions (100 K). Since earlier studies have shown that external pressure can tune SHG signal, diamond anvil cell was used to investigate the pressure dependence of SHG response.

The result of SHG measurements revealed that monochloride salt of (*S*)-2-amino-3-guanidinopropanoic acid has better optical nonlinear properties than L-arginine chloride, 3·IKDP *vs.* 0.3·IKDP. This fact can be associated with shorter carbon chain (*S*)-2-amino-3-guanidinopropanoic acid and thus closer intramolecular distance between carboxyl and guanidinium groups than in L-arginine. What is more the SHG response is more than 2 times better in 2.8 GPa pressure than in standard pressure.

Figure 1. SHG response for monochloride salt of (*S*)-2-amino-3-guanidinopropanoic acid at various pressure conditions.

Figure 1

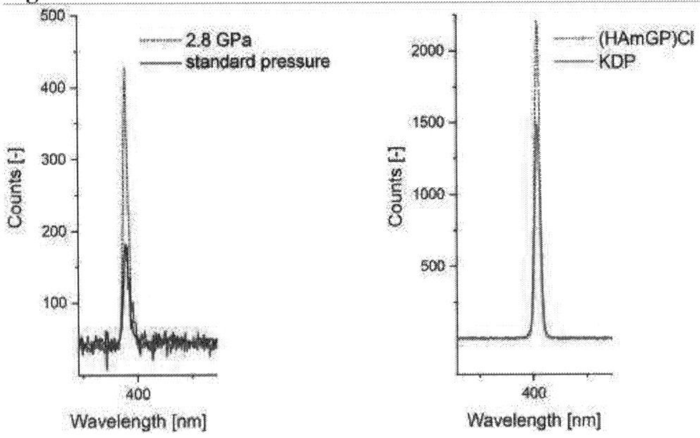

Poster

Inorganic crystal structures

P1

Diboranes(4): Synthetic Reagents with Intriguing Structures

C. Kleeberg[1], C. Borner[1], W. Drescher[1], M. T. Wiecha[1]
[1]Technische Universität Braunschweig, Institut für Anorganische und Analytische Chemie, Braunschweig, Germany

Introduction and methods

Diborane(4) derivatives such as B_2pin_2, B_2neop_2, B_2cat_2 or $B_2(NMe_2)_4$ are well established reagents in various borylation reactions.[1] Despite their ubiquitous use the molecular structures have only recently received closer attention.[2] In extension of this work we investigated the molecular structures of certain unsymmetrical amine substituted diborane(4) derivatives by single crystal X-ray diffraction (Figure).

Results and Conclusions

In the solid state catB–BMeEn comprises at 250 K individual molecules stacked along the *a* axis. Upon cooling to 100 K pairs of molecules of catB BMeEn dimerise reversibly by N···B coordination.[3]

pinB–BMeEn a liquid solidifying at –3 °C and was crystalises *in situ* on the diffractometer. At 235 K the structure comprises individual molecules of pinB–BMeEn situated on mirror planes in an orthorhombic space group *Pnma* (*Z*=4, *Z''*=½), whereas, at 100 K a lower symmetry structure in $P2_1/n$ (*Z*=4,*Z'*= 1) not exhibiting any molecular symmetry is observed, accompanied by non-merohedral twinning.[3]

enB–Bdab exhibits an extended H-bond network of individual molecules situated on a mirror plane. The structure is incommensurately modulated in a super-space group *Pnma*(00γ)0*s*0 with a modulation wave vector of [0, 0, 0.27] (*Z*=4, *Z'*=½).[4]

A detailed discussion on the individual structures and structure determinations will be given, emphasising the need for detailed, carefully conducted and comprehensive X-ray structure analysis in molecular main group chemistry and catalysis.

[1] E. C. Neeve, S. J. Geier, I. A. I. Mkhalid, S. A. Westcott, T. B. Marder, *Chem. Rev.* 2016, *116*, 9091.
[2] M. Eck, S. Würtemberger-Pietsch, A. Eichhorn, J. H. J. Berthel, R. Bertermann, U. Paul, H. Schneider, A. Friedrich, C. Kleeberg, U. Radius, T. B. Marder, *Dalton Trans.* 2017, *46*, 3661.
[3] W. Oschmann, C. Borner, C. Kleeberg, *Dalton Trans.* 2018, *47*, 5318.
[4] C. Borner, M. T. Wiecha, C. Kleeberg, *Eur. J. Inorg. Chem.* 2017, 4485.

Figure 1

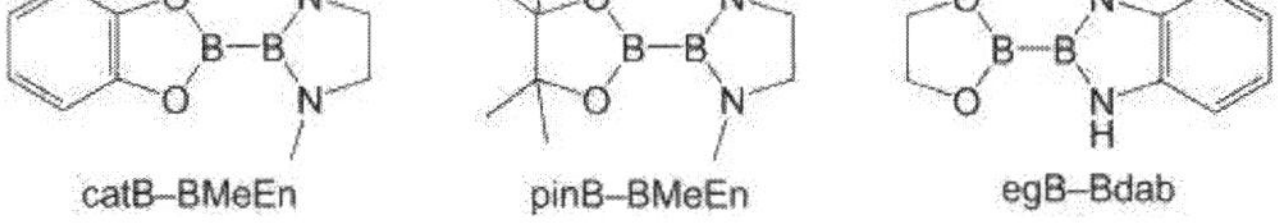

Figure: Structural formulas of the studied diborane(4) derivatives.

P2

TAAM against electron diffraction data for ionic structures

B. Gruza[1], M. L. Chodkiewicz[1], P. M. Dominiak[1]
[1]University of Warsaw, Faculty of Chemistry, Warsaw, Poland

Recently we proposed TAAM model for refinement against electron diffraction data (Gruza *et al.*, 2019). We tested the case of molecular crystals and showed that refinement statistics and obtained geometry (especially ADPs for non-hydrogen atoms), are better than in case of widely applied IAM. For further studies we have chosen α-quartz – well known, but not trivial structure with large ionic character (Tsirelson *et al.*, 1990; Fugel *et al.*, 2018).

The goal of this work is to confirm whether the TAAM model is also proper for structure of significant ionic character. For this purpose we simulated high resolution x-ray and electron diffraction structure factors. We tested possibility to define general atom type with a UBDB . Than we performed number of refinements with different resolution cut-off.

Here, we present comparison of statistics and quality of obtained geometry after IAM and TAAM refinement both against x-ray and electron diffraction data.

Support of this work by the National Centre of Science (Poland) through grant OPUS No.UMO-2017/27/B/ST4/02721 is gratefully acknowledged.

Fugel, M., Hesse, M. F., Pal, R., Beckmann, J., Jayatilaka, D., Turner, M. J., Karton, A., Bultinck, P., Chandler, G. S. & Grabowsky, S. (2018). *Chem. - A Eur. J.* **24**, 15275–15286.

Gruza, B., Chodkiewicz, M. L., Krzeszczakowska, J. & Dominiak, P. M. (2019). *ChemRxiv*.

Tsirelson, V. G., Evdokimova, O. A., Belokoneva, E. L. & Urusov, V. S. (1990). *Phys. Chem. Miner.* **17**, 275 292.

P3

$CaNa[Cr(OH)_6]$ – Synthesis, Crystal Structure, and Magnetic Properties

R. Albrecht[1], J. Hunger[1], T. Doert[1], M. Ruck[1]
[1]Technische Universität Dresden, Dresden, Germany

Introduction

Only a few chromium(III) hydroxides with alkali and alkaline earth metals are reported in literature, like the two katoite-like hydrogarnets $Ca_3[Cr(OH)_6]_2$ and $Sr_3[Cr(OH)_6]_2$[1] as sole quaternary alkaline earth representatives. Hydroxometallates are used as single-source precursors for energy and resource-saving synthesis of either known oxides or novel complex oxides by the elimination of water through mild thermal treatment.[2]

Materials and methods

Phase pure $CaNa[Cr(OH)_6]$ was obtained by solvothermal reaction in a highly concentrated sodium hydroxide solution. The structure was elucidated by single-crystal diffraction (Bruker Apex II), magnetic data were recorded on a VSM (Cryogenic Ltd).

Results

$CaNa[Cr(OH)_6]$ crystallizes in space group *R*3 as fourfold twin with the lattice parameter $a = 583.86(2)$ pm and $c = 1428.73(6)$ pm ($T = 100$ K). All atoms are located on the

Wyckoff position 3*a* (site symmetry 1). Two metal positions show a Na/Ca mixing with strong occupational preference: for *M*1 an occupation of 90(1) % Na and 10(1) % Ca was refined, whereas for *M*2 90 % Ca and 10 % Na results. The crystal structure consists of hydroxometallate layers, weakly connected by hydrogen bonds. The layers comprise a honeycomb structure consisting of $[(Ca/Na)(OH)_6]^{4-/5-}$ edge-sharing octahedral, see Figure. The voids of the honeycomb nets are filled with $[Cr(OH)_6]^{3-}$ octahedra. $CaNa[Cr(OH)_6]$ is paramagnetic in the measured temperature interval $2\ K \leq T \leq 300\ K$ with an experimental magnetic moment corresponding to a 3*d*3 spin system.

Figure: Crystal structure of $CaNa[Cr(OH)_6]$ in the *ab* plane showing the sodium-calcium honeycomb pattern, the mixed Na/Ca occupation and the hydrogen atoms are omitted for clarity.

[1] R. Albrecht, J. Hunger, T. Doert, M. Ruck, *Eur. J. Inorg. Chem.* 2019, 1398–1405.
[2] W. M. Chance, D. E. Bugaris, A. S. Sefat, H.-C. zur Loye, *Inorg. Chem.* 2013, *52*, 11723–11733.

Figure 1

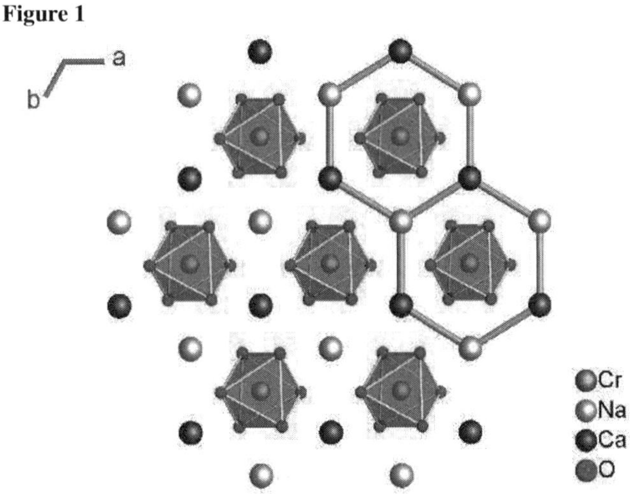

P4

Two Alkali-Metal Hexaselenidodiphosphates(IV) of Dysprosium: $LiDy[P_2Se_6]$ and $NaDy[P_2Se_6]$

B. Schulz[1,2], M. Kurz[1,2], T. Schleid[2]
[1]University of Stuttgart, Institute of Inorganic Chemistry, Stuttgart, Germany
[2]University of Stuttgart, Institute of Inorganic Chemistry, Stuttgart, Germany

The attention for selenidophosphates with rare-earth metals has increased in the last few years. Compounds with the ethane-like $[P_2Se_6]^{4-}$ anions (*d*(P–P) = 221 pm, *d*(P–Se) = 213 – 222 pm) and the structured formula *ARE*$[P_2Se_6]$ are well known for *A* = Ag, Tl and most alkali metals (*A* = Na – Cs) [1–5]. The monoclinic $NaCe[P_2Se_6]$-type structure is described in space group $P2_1/c$ with $a = 1204.0(1)$ pm, $b = 764.18(8)$ pm, $c = 1170.0(1)$ pm and $\beta = 111.269(2)°$ for $Z = 4$ [2]. Recent syntheses yielded two new representatives of the *ARE*$[P_2Se_6]$ series. Both, $LiDy[P_2Se_6]$ and $NaDy[P_2Se_6]$ crystallize in the triclinic space group with the lattice parameters shown in Table 1. In their crystal structures, only one unique Dy^{3+} cation exists, coordinated by eight selenium atoms (*d*(Dy–Se) = 290 – 328 pm). Sodium has a coordination number of *C.N.* = 6+2, while the Lithium sphere shrinks to *C.N.* = 5. The $[NaSe8]^{15-}$ polyhedra (*d*(Na–Se) = 288 – 342 pm) form strings by condensation, while the $[LiSe5]^{9-}$ polyhedra (*d*(Li–Se) = 280 – 308 pm) remain isolated due to their shorter distances to the selenium atoms (Figure 1). All hexaselenidodiphosphates(IV) with the composition *ARE*$[P_2Se_6]$ (*A* = Li and Na) were synthesized by solid-state reactions. The corresponding alkali-metal chloride (LiCl or NaCl) was mixed with the respective elemental rare-earth metal, red phosphorus and selenium. The reactants were loaded into small fused silica tubes with a molar ratio of 1:1:2:6 and tempered at 800 °C for 4 days. Amber-colored, plate-shaped single crystals of members of the *ARE*$[P_2Se_6]$ series were obtained.

[1] L. M. Schoop, R. Eger, J. Nuss, F. Pielnhofer, B. V. Lotsch, *Z. Anorg. Allg. Chem.* 2017, *643*, 1818.
[2] J. A. Aitken, M. Evain, L. Iordanidis, M. G. Kanatzidis, *Inorg.Chem.* 2002, *41*, 180.
[3] J. M. Knaust, P. K. Dorhout, *J. Chem. Crystallogr.* 2005, *36*, 217.
[4] J. Garin, E Parthé, *Acta Crystallogr.* 1972, *B 28*, 3672.
[5] A. Pfitzner, S. Seidelmayer, *Z. Anorg. Allg. Chem.* 2009, *635*, 704.

Figure 1

Figure 1. Crystal structures of $LiDy[P_2Se_6]$ (*top*) and $NaDy[P_2Se_6]$ (*bottom*) as viewed along [010].

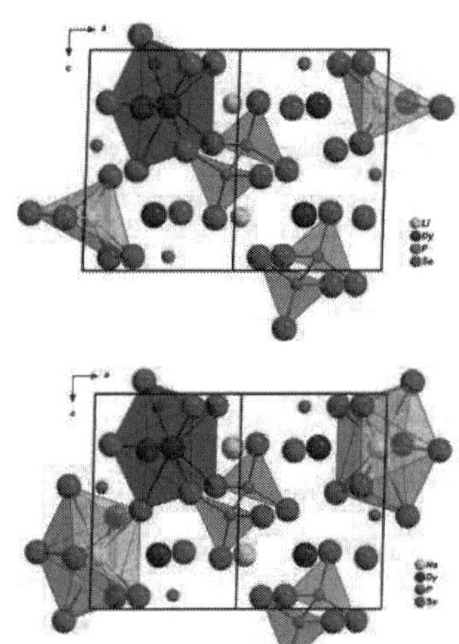

Figure 2

Table 1. Crystal data for $LiDy[P_2Se_6]$ and $NaDy[P_2Se_6]$.

Compound	$LiDy[P_2Se_6]$	$NaDy[P_2Se_6]$
space group	$P\bar{1}$	$P\bar{1}$
a/pm	667.84(5)	672.54(5)
b/pm	732.03(5)	748.16(5)
c/pm	958.62(7)	965.21(7)
α/°	90.721(2)	90.498(2)
β/°	92.156(2)	91.737(2)
γ/°	90.718(2)	90.449(2)
Z	2	2
Calculated density (D_x in g/cm³)	5.00	4.93
Molar volume (V_m in cm³/mol)	140.98	146.15
Diffractometer	κ-CCD (Bruker-Nonius)	κ-CCD (Bruker-Nonius)
Wavelength	Mo-Kα: λ = 71.07 pm	Mo-Kα: λ = 71.07 pm
Diffractometer limit ($2\Theta_{max}$ in deg)	55	55
Index range ($\pm h_{max}$, $\pm k_{max}$, $\pm l_{max}$)	8 / 9 / 12	8 / 9 / 12
Number of e⁻ per unit cell (F(000))	628	622
Absorption coefficient (μ in mm⁻¹)	31.61	30.49
Collected / unique reflections	17019 / 2143	7781 / 2244
Data set residuals (R_{int} / R_σ)	0.105 / 0.049	0.091 / 0.042
Structure residuals (R_1 / wR_2)	0.042 / 0.107	0.063 / 0.148
Goodness of fit (GooF)	1.094	1.102
Residual electron density (max. / min. in 10^6 pm⁻³)	1.91 / −1.87	1.85 / −2.16
CSD number	1961901	1961902

P5

Single Crystals of Nb_5Se_4 with Ti_5Te_4-type Structure

C. Buyer[1], T. Schleid[1]
[1]University of Stuttgart, Institute of Inorganic Chemistry, Stuttgart, Germany

The crystal structure of Nb_5Se_4 was first reported in 1963 by Selte and Kjekhus based on the powder pattern of a sample obtained by heating a mixture of niobium and selenium in a evacuated and sealed silica tube at about 1000 °C [1]. Thus Nb_5Se_4 crystallizes with the crystal structure of Ti_5Te_4 [2] in the tetragonal space group *I*4/*m*. In 2015 Klimczuk et al. synthezised polycrystalline Nb_5Se_4 and measured superconductivity for this compound [3].

We now encountered black, needle-shaped, single crystals of Nb_5Se_4 (CSD-1962241) with a = 986.79(8) and c = 344.68(3) pm exhibiting diameters of about 0.01 mm and 3 mm length. The target of the synthesis was to find single crystals of YFSe polymorphs analogous to HoFSe [4]. Therefore a mixture of yttrium, yttrium trifluoride and selenium was mixed with lead difluoride as a fluxing agent in a sealed niobium ampoule, heated to 860 °C for 5 days and cooled down with 4 °C per hour. An unexpected reaction between the niobium container and selenium led the formation of Nb_5Se_4 single crystals and their stoichiometric composition was verified by WDX measurements on an electron-beam X-ray microprobe.

The crystal structure contains 1D-$[(Nb2)_4(Nb1)_{2/2}]$ chains of trans-vertex-sharing octahedra of niobium running along the c axis. Shared niobium is square-planar surrounded by selenium at a distance of 284 pm and has a total coordination number of 12 (4 × Se, 8 × Nb). The Nb—Nb distances are with 287 pm nearly the same as those between Nb1 and Se, while the five Nb2—Se distances between 261 and 267 pm appear much shorter.

[1] K. Selte, A. Kjekshus: *Acta Chem. Scand.* 1963, 17, 2560–2574.
[2] F. Grønvold, A. Kjekshus, F. Raaum: *Acta Crystallogr.* 1961, 14, 930–934.
[3] T. Klimczuk, K. Baroudi, J. W. Krizan, A. L. Kozub, R. J. Cava: *J. Alloys Compd.* 2015, 649, 906-909.
[4] D. D. Zimmermann, Th. Schleid: *Z. Kristallogr.* 2014, *Suppl.* 34, 139–139.

Figure 1

Table 1. Atomic coordinates and equivalent isotropic displacement coefficients (U_{eq}/pm²) for Nb_5Se_4

Atom	Site	x/a	y/b	z/c	U_{eq}
Nb1	2*a*	0	0	0	86(4)
Nb2	8*h*	0.30514(9)	0.37252(9)	0	110(3)
Se	8*h*	0.05713(9)	0.28159(9)	0	108(4)

Figure 2

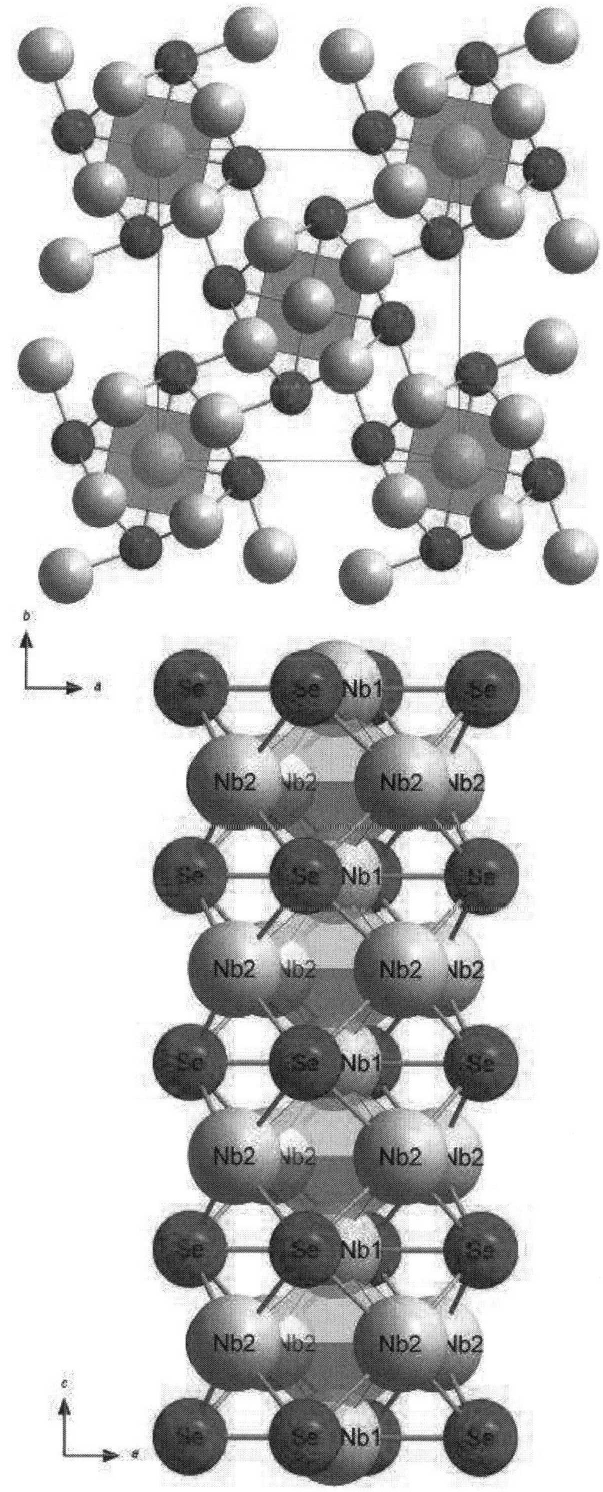

Figure 1. Tetragonal unit cell of Nb_5Se_4 as (001) projection (top) and selenium-mantled $^1_\infty[(Nb2)_4(Nb1)_{2/2}]$ chain running along [001] (bottom).

P6

Coordination properties of diethylenetriamine in relation to zinc phthalocyanine

J. Janczak[1]
[1]Institute of Low Temperature and Structure Research, Polish Academy of Sciences, Crystallography, Wrocław, Poland

In the contribution the coordination properties of diethylenetriamine (DETA) as a ligand containing three donor nitrogen atoms, two terminal primary amino groups and one secondary central amino group, with respect to divalent zinc ion in ZnPc were examined. The ZnPcDETA complex in the powder form was obtained by the solvothermal reaction of ZnPc with DETA. Recrystallization of the crude product from 3,4-lutidine yields non-centrosymmetric monoclinic solvated crystals $(ZnPcDETA)_3$.3,4-lut (**1**) with the space group of Cc. X-ray single crystal analysis of **1** shows that the DETA coordinates *via* terminal amine group to the divalent zinc ion of a planar ZnPc molecule in axial position. Interaction of N atom containing a lone electron pair with zinc ion of ZnPc and the formation of the Zn-N coordination bond leads to deviation of the Zn from the N4-isoindole plane of Pc by ~0.5 Å toward the N-atom of the axial DETA ligand as well as to saucer-shaped distortion of the Pc macrocycle. Arrangement of **1** in crystal is determined by the van der Waals forces and by the weak N-H...N hydrogen bonds. The lack of the π... π interaction between the phthalocyaninate(2-) macrocycles in the present structure is clearly evidenced by the Hirshfeld surface analysis and leads to significantly increasing its solubility in most common solvents. This feature together with the strong absorption in the therapeutic window within the range of 600-900 nm makes its a potentially good photosensitizer. DFT calculations performed for the ZnPc-derivatives with DETA, coordinated *via* terminal or central amine group of DETA as well as for the bridged complex show the possibility to obtain all three of these complexes. UV-Vis absorption spectra of **1** in solutions as well as the diffuse reflectance spectroscopy (DSR) supported by the TD-DFT calculations were used for the characterization of the spectroscopic properties. SHG efficiency of **1** is ~10 % in relation to that of KDP.

P7

Disorder in $Ca_3RE_2(BO_3)_4$ (RE = Nd, Gd) structure

K. M. Kosyl[1,2], A. Gągor[3], W. Paszkowicz[1], A. N. Shekhovtsov[4], M. B. Kosmyna[4], D. Trzybiński[5], K. Woźniak[5]
[1]Institue of Physics PAS, Warsaw, Poland
[2]European Synchrotron Radiation Facility (ESRF), Grenoble, France
[3]Institute of Low Temperature and Structure Research, Polish Academy of Sciences, Wrocław, Poland
[4]Institute for Single Crystal NASU, Kharkiv, Ukraine
[5]University of Warsaw, Department of Chemistry, Warsaw, Poland

The borates of $A_3RE_2(BO_3)_4$ formula (A=Ca/Sr/Ba; RE=a rare earth or Bi; space group *Pnma*) are materials potentially applicable in optoelectronics (e.g. for 1.5 μm solid state lasers). Compounds from this family can be synthesised by solid state technique and grown by the Czochralski method, allowing uniform doping of the host by other rare earths.

The characteristic feature of the structure is sharing three cationic sites by A and RE atoms, surrounded by 8-9 oxygen atoms. Between as-created irregular polyhedra, three boron sites are present, in the centres of triangles formed by three neighbouring oxygen ions.

During the powder XRD studies on calcium borates from this family, we noticed anomalously large atomic displacements of boron atoms. As the structure is quite complex, we decided to perform single crystal XRD for two representative members of this family. The $Ca_3RE_2(BO_3)_4$ (RE=Nd, Gd) crystals were grown by Czochralski method (for growth details see [1]). A small piece of single crystal was extracted and measured at 300 (RE=Gd) and 100 K (RE=Nd, Gd).

$Ca_3RE_2(BO_3)_4$ (RE=Nd, Gd) are found to be isostructural with the Sr- and Ba-based borates of the same formula, although some new features can be noticed when going into more detail – e.g. all the three cationic sites exhibit splitting into two separate Ca and RE positions. Additionally, two over three borate groups are disordered. The character of the disorder is different for each case – at one of them oxygen atoms are located at two well-distinguished configurations, at the latter one an electron density of oxygen ions is more blurred. Applying appropriate modifications into a structure description results with much more reasonable atomic displacement parameters.

In summary, the structure of $Ca_3RE_2(BO_3)_4$ (RE=Nd, Gd) was solved. A new insight into disorder of the $A_3RE_2(BO_3)_4$-type materials is proposed, on the basis of single-crystal XRD experiments.

[1] L. V. Gudzenko et al. (2017) *Crystals*, *7*(3), 88.

P8

Structure refinement for $Ca_{10}M_{0.5}(VO_4)_7$ M=Co, Cu: A powder X-ray diffraction study

H. S. Rahimi Mosafer[1], W. Paszkowicz[1], M. Berkowski[1]
[1]institute of physics, polish academy of science, Warsaw, Poland

Calcium orthovanadate and orthophosphate $Ca_3(XO_4)_2$, X = P or V, is known to crystallize in R3c space group. By substitution of alkali or rare earth to this material, synthetic materials related to those of whitlockite group minerals are obtained. For some of them, applications in different areas are considered.

In this study, we investigate the structural properties of $Ca_{10}M_{0.5}(VO_4)_7$ where M are cobalt and copper. The insight of the structural properties of these materials, not reported until now, is therefore important for their full understanding and characterization.

Solid solution of new materials, $Ca_{10}M_{0.5}(VO_4)_7$, where M= Cu, Co, were synthesized by solid state reaction.

Powder diffraction measurement was performed by Philips X"Pert Pro Alpha1 diffractometer with Bragg-Brentano geometry.

Phase analysis of $Ca_{10}M_{0.5}(VO_4)_7$ has shown that samples crystalize in R3c group. Diffraction peaks are assigned to those of other whitlockite–type material.

The starting model for refinement of these material was $Ca_{1.5}Sr_{1.5}(VO_4)_7$ [1]. Lattice parameters decrease by substitution of M [M=Cu, Co]. The ideal stoichiometric ratio in the unit cell is 4.76%. These refinements bring 4.62% for cobalt sample and 3.64% for copper sample.

Among the five available Ca sites, only Ca(5) is found to be occupied by the transition metal, this finding is in line with observation for related phosphates [2]. Moreover, this coincides with analysis of interatomic distances.

Structure properties of $Ca_{10}Cu_{0.5}(VO_4)_7$ and $Ca_{10}Co_{0.5}(VO_4)_7$ have been determined. A shrinkage in lattice parameters is caused by entering of small amount of copper and cobalt into calcium vanadate lattice. Rietveld refinements demonstrate that copper and cobalt enter to M5 site, what influences the interatomic distances.

[1] A. A. Belik *et al.*, *Mater. Res. Bull.*, vol. 36, no. 10, pp. 1873–1880, 2001.
[2] A. Altomare *et al.*, *Crystals*, vol. 9, no. 6, pp. 1–17, 2019.

P9

Uncontrolled Synthesis and Crystal Structure of $La_5O_4Cl_3[TeO_3]_2$

P. Russ[1], S. Greiner[1], T. Schleid[1]
[1]University of Stuttgart, Institute of Inorganic Chemistry, Stuttgart, Germany

As by product from the synthesis $Na_2La_3Cl_3[TeO_3]_4$ [1—3] single crystals of $La_5O_4Cl_3[TeO_3]_2$ (CSD-434006) could be obtained as secondary phase. For the synthesis a mixture of La_2O_3, TeO_2 and NaCl as flux and reactant was prepared in a molar ratio of 2:4:6 and given into an evacuated fused silica glass ampoule. Within 10 hours it was heated up to 950 °C. This temperature was held for 5 days and then the furnace was cooled down to 500 °C with 5 °C/h, before it was switched off.

$La_5O_4Cl_3[TeO_3]_2$ crystallizes in the monoclinic space group *C2/m* with the lattice parameters a = 1304.45(9) pm, b = 572.96(4) pm, c = 1021.11(8) pm, β = 91.770(3)° for Z = 2, just like $Gd_5O_4Br_3[SeO_3]_2$ [4]. In its crystal structure there are three different La^{3+}-cation positions. The first one is coordinated by eight oxygen atoms in a square $[(La1)O_8]^{13-}$ prism. There are two edges coordinated by ψ^1-tetrahedral $[TeO_3]_2$— units and the remaining four oxide anions are not bound to tellurium. The second La^{3+} cation is surrounded trigonally by six oxygen atoms, which belong to two terminal $[TeO_3]_2$— anions. The other four oxide anions are unbound to tellurium and additionally a rectangular face is capped by a chloride anion resulting in a $[(La2)O_6Cl]^{10-}$ prism. The third La^{3+} cation is in an environment of five oxygen atoms, which belong to three terminal $[TeO_3]^{2-}$ units and two oxide anions. In addition, the La^{3+} cation is surrounded by three chloride anions and therefore the result is a square $[(La3)O_5Cl_3]^{10-}$ antiprism.

In the three-dimensional network, it can be noted that a layered partial structure propagates parallel to the (001) plane with empty channels extending along [010] (Figure 1). The non-tellurium bonded oxide anions are tetrahedrally surrounded by four La^{3+} cations. The $[OLa_4]^{10+}$ tetrahedra share common edges to form 1D-$\{[O_2La_3]^{5+}\}$ double chains. These double chains are further condensed via common corners to 2D-$\{[O_4La_5]^{7+}\}$ layers.

Figure 1

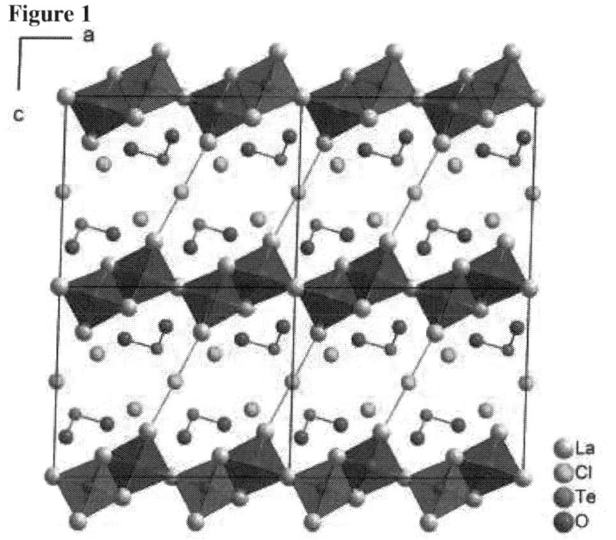

Figure 1. Layer-like arrangement of linked $[OLa_4]^{10+}$ tetrahedra in the crystal structure of $La_5O_4Cl_3[TeO_3]_2$.

Figure 2

[1] S. Zitzer, *Doctoral Dissertation*, Universität Stuttgart, **2012**.
[2] S. Greiner, *Doctoral Dissertation*, Universität Stuttgart, **2018**.
[3] D. O. Charkin, S. Zitzer, S. G. Dorofeev, A. V. Olenev, P. S. Berdonosov, Th. Schleid, V. A. Dolgikh, *Z. Anorg. Allg. Chem.* **2017**, 643, 1654.
[4] D.-H. Kang, J. Wontcheu, Th. Schleid, *Solid State Sci.* **2009**, 11, 299.

Figure 2. References

P10

$Nd_{1.333}[P_2Se_6]$: Neodymium(III) Hexaselenidodiphosphate(IV) with Cation-Deficient $NaYb[P_2S_6]$-Type Crystal Structure

M. Kurz[1], B. Schulz[1], T. Schleid[1]
[1]University of Stuttgart, Institute of Inorganic Chemistry, Stuttgart, Germany

Despite their fascinating physical and chemical properties, the so far discovered compounds with the formula $ARE[P_2Se_6]$ (A = Na – Cs, RE = Y, La – Pr, Sm – Tm [1–4]) show a large structural variety and even some stoichiometric flexibility [5]. The already known $Ce_{1.333}[P_2Se_6]$ crystallizes in the monoclinic space group $P2_1/c$ with the cation-deficient $NaCe[P_2Se_6]$-type structure [2], but the new neodymium selenidophosphate $Nd_{1.333}[P_2Se_6]$ crystallizes with the $NaYb[P_2S_6]$-type analogue [6]. For this structure, the triclinic space group P-1 with the lattice parameters a = 682.41(5) pm, b = 757.98(6) pm, c = 961.03(7) pm, α = 90.176(2)°, β = 91.789(2)° and γ = 90.108(2)° is found for Z = 2.

The crystal structure of $Nd_{1.333}[P_2Se_6]$ (Figure 1) contains two crystallographically different Nd^{3+} cations, both surrounded by eight selenium atoms (Figure 2, *left*). Bicapped trigonal prisms $[NdSe_8]^{13-}$ (d(Nd–Se) = 300 – 337 pm) are connected via Se···Se edges to form single strings. While one neodymium position is fully occupied, the second one is only occupied by one third, resulting in the formula $Nd_{1.333}[P_2Se_6]$. All selenium atoms belong to ethane-like $[P_2Se_6]_{4-}$ anions with d(P–P) = 221 – 224 pm and d(P–Se) = 217 – 219 pm (Figure 2, *right*).

To synthesize the yellow single crystals of $Nd_{1.333}[P_2Se_6]$, neodymium powder was mixed with red phosphorus, selenium and sodium chloride (NaCl) as flux in a molar ratio of 1:2:8:1. The synthesis was carried out in evacuated fused silica ampoules at 800 °C for 4 days.

[1] L. M. Schoop, R. Eger, J. Nuss, F. Pielnhofer, B. V. Lotsch, *Z. Anorg. Allg. Chem.* 2017, *643*, 1818.
[2] J. A. Aitken, M. Evain, L. Iordanidis, M. G. Kanatzidis, *Inorg. Chem.* 2002, *41*, 180.
[3] J. M. Knaust, P. K. Dorhout, *J. Chem. Crystallogr.* 2006, *36*, 217.
[4] J. Garin, E. Parthé, *Acta Crystallogr.* 1972, *B 28*, 3672.
[5] J. H. Chen, P. K. Dorhout, J. Ostenson, *Inorg. Chem.* 1996, *35*, 5627.
[6] E. Y. Goh, E. J. Kim, S. J. Kim, *J. Solid State Chem.* 2001, *160*, 195.

Figure 1

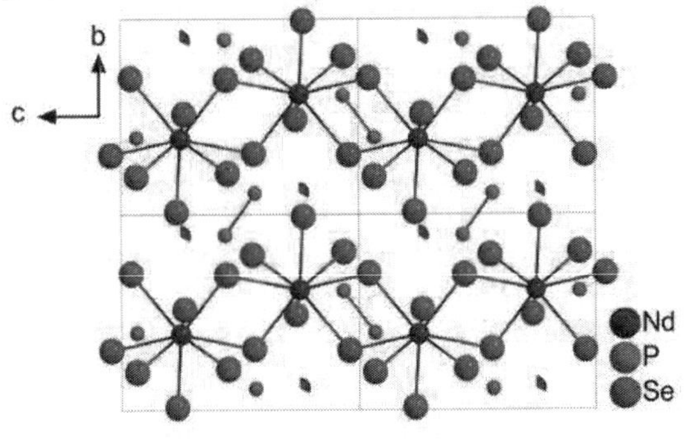

Figure 1. Crystal structure of $Nd_{1.333}[P_2Se_6]$ as viewed along [100] with highlighted Nd–Se bonds.

Figure 2

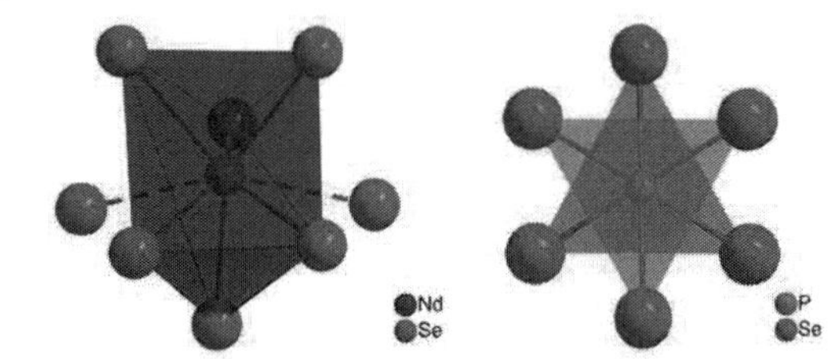

Figure 2. Bicapped trigonal prisms $[NdSe_8]^{13-}$ (*left*) and the staggered conformation of the ethane-like $[P_2Se_6]^{4-}$ anions as Newman projection along the P–P bond (*right*) in $Nd_{1.333}[P_2Se_6]$.

P11

Synthesis and Crystal Structure of $Tl_2[B_{10}Cl_{10}] \cdot 2\ H_2O$

K. Bareiß[1], T. Schleid[1]
[1]University of Stuttgart, Institute of Inorganic Chemistry, Stuttgart, Germany

The previously reported crystal structure of the green $Cu(H_2O)_4[B_{10}Cl_{10}] \cdot 5\ H_2O$ by *Kleeberg* and *Schleid* [1] led to the preparation of further perchlorinated B_{10}-cluster analogues, such as $Tl_2[B_{10}Cl_{10}] \cdot 2\ H_2O$. We now present the synthesis and structural characterization of $Tl_2[B_{10}Cl_{10}] \cdot 2\ H_2O$ (CSD-1965968) crystallizing isotypically with $Cs_2[B_{10}Br_{10}] \cdot 2\ H_2O$ [2] in the monoclinic space group *C*2/*c* with $a = 3018.53(19)$ pm, $b = 1145.16(7)$ pm, $c = 1282.31(8)$ pm and $\beta = 106.712(3)°$ for Z = 8. The $[B_{10}Cl_{10}]^{2-}$-cluster anions feature typical B–B bond lengths of 167 – 168 pm to the lids and 182 – 184 pm inside the square antiprism as well as B–Cl distances from 178 to 181 pm, which nicely fit the values of triclinic $Cu(H_2O)_4[B_{10}Cl_{10}] \cdot 5\ H_2O$ [1]. The three crystallographically different Tl^+-cation sites, of which two are partially occupied, lead to a total amount of only two per formula unit. The fully occupied Tl1 site is coordinated by seven Cl^- anions and two oxygen atoms from water molecules forming an irregular polyhedron with an O1···O1 edge (Figure 1, left). Both partially occupied thallium sites, Tl2 (*s.o.f.* = 0.62) and Tl3 (*s.o.f.* = 0.38), are coordinated by six Cl^- anions and one formally shared water molecule represented by O2, but for Tl3 one additional chloride ligand can be considered (Figure 1, right). Owing to the short distance of 343 pm between the Tl2 and Tl3 positions and the fact that both rates of occupation sum up to one, it can be predicted that only one site at a time would be occupied.

In order to prepare $Tl_2[B_{10}Cl_{10}] \cdot 2\ H_2O$ thallium(I) carbonate ($Tl_2[CO_3]$) is neutralized by using an aqueous acid solution of $(H_3O)_2[B_{10}Cl_{10}]$. The reaction mixture was filtrated and the solution slowly evaporated isothermally to obtain colorless crystals of $Tl_2[B_{10}Cl_{10}] \cdot 2\ H_2O$ from the mother liquor.

[1] F. M. Kleeberg, Th. Schleid, *Z. Kristallogr.* 2017, *Suppl. 37*, 107–107.
[2] W. Einholz, K. Vaas, C. Wieloch, B. Speiser, T. Wizemann, M. Ströbele, H.-J. Meyer, *Z. Anorg. Allg. Chem.* 2002, *628*, 258–268.

Figure 1

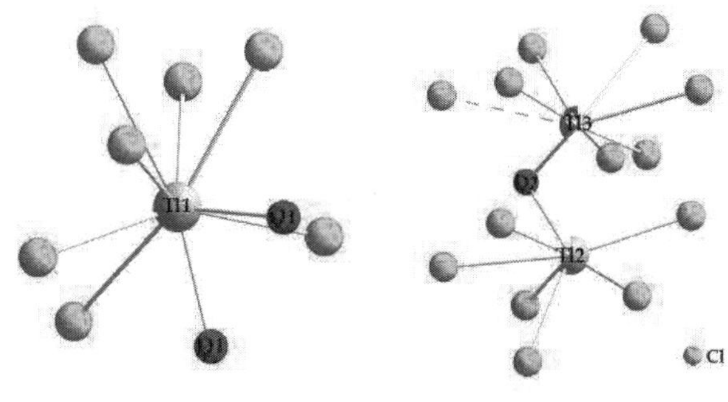

Figure 1. First coordination spheres of Tl1 (left) and of Tl2 and Tl3 (right) in the crystal structure of $Tl_2[B_{10}Cl_{10}] \cdot 2\ H_2O$.

Figure 2

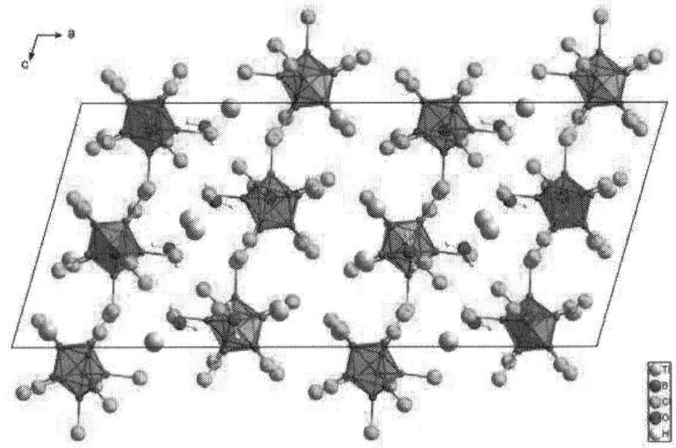

Figure 2. View at the crystal structure of $Tl_2[B_{10}Cl_{10}] \cdot 2\ H_2O$ along [010].

P12

The Structure of Yttrium Hydride Telluride YHTe from Single-Crystal X-Ray Diffraction Data

J. L. Hoslauer[1], M. Folchnandt[1], T. Schleid[1]
[1]University of Stuttgart, Institute of Inorganic Chemistry, Stuttgart, Germany

In 1961, Carter reported of several hydride chalcogenides, for which he suggested a cubic crystal structure for the composition LnHxO ($0.56 \leq x \leq 1.41$) and a tetragonal one for LnH$_x$Te ($0.91 \leq x \leq 1.41$) [1]. Later it could be shown that hydride oxides of the composition LnHO crystallize in a tetragonal ordered fluorite-type structure for *Ln* = La, Ce and Pr [2], while for *Ln* = Sm, Gd – Er an anion-disordered cubic fluorite-type structure is found [3, 4]. On the contrary, the hydride selenides *Ln*HSe show a hexagonal 2H-type structure (space group: $P6_3/mmc$) for *Ln* = La – Nd as well as a hexagonal 1H-type structure (space group: *P*-6*m*2) for *Ln* = Y, Gd – Tm, Lu [5]. In here, we report on the tellurides *Ln*HTe (*Ln* = La – Nd, Gd – Er) [6], adding a crystal structure determination for the yttrium hydride telluride YHTe.

Yttrium metal pieces were hydrogenated at 450 °C for 10 h at a starting pressure of 500 mbar to yield the corresponding dihydride YH_2, which was further reacted with elemental tellurium at 700 °C for 7 days with CsCl as fluxing agent. The components were sealed in cleaned arc-welded tantalum capsules, which were further jacketed with fused silica ampoules to prevent oxidation. The product mixture was washed with demineralized water and single crystals of YHTe were selected to determine the crystal structure.

YHTe (space group: *P*-6*m*2) is best described as filled WC-type structure. The Y^{3+} cations and the Te^{2-} anions form a hexagonal primitive packing each, in which both ions fill the trigonal prismatic voids of the other. For Y^{3+} this coordination sphere is further increased by three capping H^- anions (Figure 1). The interatomic distances of 306 pm for *d*(Y–Te) and 231 pm for *d*(Y–H) are in good agreement with those of the sesquitelluride Y_2Te_3 (*d* = 305 – 308 pm) [7] and the trihydride YH_3 (*d* = 224 – 247 pm) [8].

Figure 1

Table 1a. Crystallographic data for YHTe and their Determination.

Formula	YHTe
Space group	$P\bar{6}m2$ (no. 187)
Crystal system	hexagonal
a / pm	402.44(2)
c / pm	414.21(3)
c / *a*	1.029
Number of formula units, *Z*	1
Calculated density, D_x/g · cm^{-3}	6.52
Molar volume, V_m/cm^{-3} · mol^{-1}	34.986(5)
Diffractometer	Four-circle AED 2 (Siemens-Stoe)
Radiation; Wavelength / pm	Mo-K_α, 71.07
F(000) / e	92
Range (in *θ* / deg)	1 – 40
Absorption, *μ* / mm^{-1}	37.16
Total number of reflections	1455
Number of symmetrically independent reflections	177
R_{int}	0.031
R	0.016
R_w	0.015
Structure solution and refinement	SHELXS/L-97
Extinction, *g* · 10^3	72(2)

Table 1b. Fractional atomic coordinates as well as isotropic equivalent and anisotropic thermal displacement parameters (U_{ij}[a] / U_{eq}[b] in pm^2) for YHTe ($U_{22} = U_{11}$, $U_{12} = {}^1/_2\, U_{11}$, $U_{23} = U_{13} = 0$ due to space-group symmetry).

Atom	Site	*x*/*a*	*y*/*b*	*z*/*c*	U_{11}	U_{33}	U_{12}	U_{eq}
Y	1*f*	$^2/_3$	$^1/_3$	$^1/_2$	141(2)	128(4)	71(1)	136(3)
H	1*d*	$^1/_3$	$^2/_3$	$^1/_2$	-	-	-	477[c]
Te	1*a*	0	0	0	115(1)	145(2)	57(1)	125(3)

a) $U_{ij} = \exp[-2\pi^2 (a^{*2}h^2 U_{11} + \ldots + b^* c^* k l\, 2 U_{23} + \ldots)]$,

b) $U_{eq} = {}^1/_3 [U_{33} + {}^4/_3 (U_{11} + U_{22} - U_{12})]$,

c) The isotropic displacement parameter U_{iso} of the hydrogen atom was constrained to the parameter U_{eq} of the corresponding metal atom with a factor of 3.5.

Figure 2

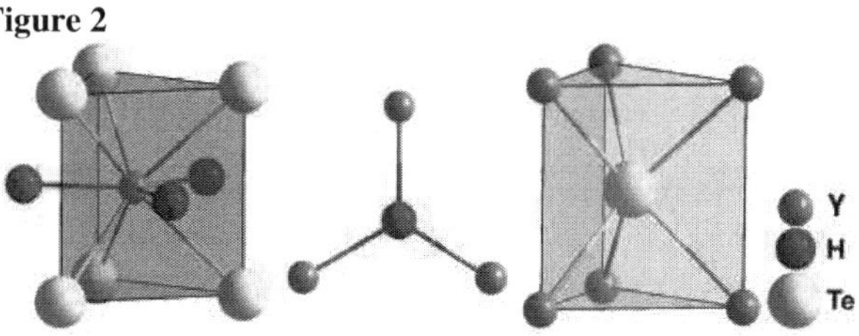

Figure 1. Coordination polyhedra in the crystal structure of YHTe.

[1] F. L. Carter, *Proc. 2nd Rare-Earth Res. Conf.* Denver CO (USA) **1961**, p. 311.

[2] J. F. Brice, *J. Solid State Chem.* **1984**, *53*, 44.

[3] H. Yamashita, T. Broux, Y. Kobayashi, F. Takeiri, H. Ubukata, T. Zhu, M. A. Hayward, K. Fujii, M. Yashima, K. Shitara, A. Kuwabara, T. Murakami, H. Hageyama, *J. Am. Chem. Soc.* **2018**, *140*, 11170.

[4] N. Zapp, H. Kohlmann, *Z. Naturforsch.* **2018**, *73 b*, 535.

[5] Th. Schleid, M. Folchnandt, *Z. Anorg. Allg. Chem.* **1996**, *622*, 455.

[6] M. Folchnandt, D. Rudolph, J.-L. Hoslauer, Th. Schleid, *Z. Naturforsch.* **2019**, *74 b*, 513.

[7] J. Flahaut, P. Laruelle, M. P. Pardo, M. Guittard, *Bull. Soc. Chim. Fr.* **1965**, 1399.

[8] W. E. Wallace, *J. Phys. (Paris)*, **1964**, *25*, 454.

P13

Phase transitions in crystals of tertiary ammonium salts with reo4- ions

M. Krawczyk[1]

[1]Wrocław University of Science and Technology, Institute of Experimental Physics, Department of Physics and Astronomy, Wrocław, Poland

Salts and coordination compounds containing transition metals are extensively studied due to the fact they may exhibit phase transitions connected to their ferroic properties. In search of new ferroic materials, studies focused on salts of rhenium(VII) with N-heterocyclic compounds, such as hexamethylenetetramine were undertaken. On the basis of thermal analysis and single crystal X-ray diffraction experiments it was found that for the crystal comprised of hexamethylenetetrammonium and ReO_4^- ions several phase transitions can be observed. Upon cooling the two phase transitions at about 240 and 233 K occur. Afterwards, the studied crystal undergoes the phase transitions at about 263 and 284 K on heating. The presented studies concern the analysis of structural changes in the crystal in relation to the phase transitions.

P14

HKLF5Tools: Small Helper in Refinement of Non-Merohedral Twins

S. Ivlev[1], M. Conrad[1], F. Kraus[1]

[1]Philipps-Universität Marburg, Fachbereich Chemie, Marburg, Germany

Twinning is one of the common reasons that prevents a crystal structure from being solved and refined routinely. Practically any structure refinement software can successfully handle twins by merohedry, pseudomerohedry, and partial merohedry, provided a proper twin operation is given. In case of non-merohedral twinning, in which the reciprocal lattices of the twin components do not overlap systematically, one standard procedure is using the reflection file in HKLF5 format. [1,2] However, it may not always be straightforward to carry out a proper refinement. In general, when a crystal consisting of non-merohedrally twinned domains is recognized, the crystallographer may need additional software tools which are not implemented in the primary structure refinement program.

Here we will present the HKLF5Tools program and its latest developments. This program provides the crystallographer with a set of tools for structure refinements based on non-merohedrally twinned crystals. The software shows statistical information on reflections belonging to each twin component, can add, delete or rename labels of twin components, and merge reflection data. Also, the software can convert ShelXL FCF files into HKL reflection files so that refinement against detwinned data becomes possible.

References
[1] G. M. Sheldrick. *Acta Crystallogr., Sect. C: Struct. Chem.* 2015, *71* (*1*), 3–8.
[2] R. Herbst-Irmer, G. M. Sheldrick. *Acta Crystallogr., Sect. B: Struct. Sci.* 2002, *58* (*3*), 477–481.
[3] S. I. Ivlev, M. Conrad, F. Kraus. *Z. Kristallogr. - Cryst. Mater.*, 2019, *234* (*6*), 415-418.

P15
Lithium Hypodiphosphates

V. Kinzhybalo[1]
[1]Institute of Low Temperature and Structure Research, Polish Academy of Sciences, Wrocław, Poland

In recent years hypodiphosphoric acid, $(HO)_2OP–PO(OH)_2$, and its salts have attracted attention due to the discovery of ferroelectric properties in diammonium dihydrogen hypodiphosphate, $(NH_4)_2H_2P_2O_6$ [1]. Ferroelectric properties turned out to depend on the hydrogen bonding network formed in the crystal. The systematic study on inorganic salts of $H_4P_2O_6$ was undertaken by several scientific groups [2-5]. Up to now one of the best studied group of salts is alkali metals hypodiphosphates. Nevertheless lithium slats are represented by only two examples: lithium dihydrogen hypodiphosphate dihydrate, $Li_2H_2P_2O_6{\cdot}2H_2O$ [2] and lithium hypodiphosphate hexahydrate, $Li_4P_2O_6{\cdot}6H_2O$, for which experimental electron density distribution was studied [3]. In general, inorganic lithium salts are used in nonlinear optics and lithium-ion batteries. Therefore a systematic study on lithium hypodiphosphates was undertaken. In result several substances of the following composition were obtained and structurally characterized: $LiH_3P_2O_6$ (1), $Li_2H_2P_2O_6$ (2), another form of $Li_2H_2P_2O_6{\cdot}2H_2O$ (3), $Li_3NaP_2O_6{\cdot}6H_2O$ (4) and $Li_2K_2P_2O_6{\cdot}8H_2O$ (5).
Compounds 1-5 were prepared by addition of appropriate solid alkali metals hydroxides to an aqueous solution of hypodiphosphoric acid.
In all of the substances lithium cation possess tetrahedral environment. Hypodiphosphate anions coordinate to 4, 6 or 8 lithium cations. In acid salts hypodiphosphate anions form anti-electrostatic hydrogen-bonded networks with each other.

Literature
[1] P. Szklarz, M. Chański, K. Ślepokura, T. Lis, *Chem. Mater.* 23 (2011) 1082.
[2] P. Wu, M. Gjikaj, *Joint Meeting: DGK, DMG, ÖMG* (2011) P10-P24, 101.
[3] V. Kinzhybalo, A. Mermer, T. Lis, P. Starynowicz, *Acta Cryst.* B69 (2013) 344.
[4] B. Szafranowska, K. Ślepokura, T. Lis, *Acta Cryst.* C68 (2012) i71.
[5] M. Gjikaj, P. Wu, W. Brockner, *Z. Anorg. Allg. Chem.* 638 (2012) 2144.

P16
The crystal structure of trinuclear rhenium coordination compound with Re3(O)(NO)6(pz)6]Cl·2CH3CN formula.

M. Siczek[1]
[1]Wrocław University of Science and Technology, Faculty of Chemistry , Wrocław, Poland

Polynuclear rhenium complexes, despite numerous synthesis attempts, still belong to one of the least known rhenium compounds. Synthesis of most rhenium clusters often requires the use of high temperature and high pressure[1]. Therefore it is challenging to find a new synthesis methods that allow to obtain a new rhenium cluster in a mild conditions.

Herein a new method of synthesis of trinuclear rhenium coordination compounds by reductive nitrosylation of perrhenate is presented. The reaction in basic condition between perrhenate and hydroxylamine led to trinuclear nitrosyl compound. Further reaction in acetonitrile with hydrochloric acid and pyrazole gave crystals of [Re3(O)(NO)6(pz)6]Cl·2CH3CN (**1**).

The compound crystallizes in the triclinic *P* space group, The cation [Re3(O)(NO)6(pz)6]+ consists of three rhenium atoms bridged with three nitrosyl ligands and one oxygen atom μ3-O. Six nitrosyl ligands (three bridging and three terminal) and oxygen atom coordinate with each of the rhenium atoms

In summary this study reports coordination-chemistry and synthesis method of a new class of rhenium compounds. To the best of our knowledge, this is the first report of trinuclear rhenium compound with nitrosyl ligand.

[1] Gray, T. G. Coord. Chem. Rev. 2003, 243, 213– 235.

Micro- and nanocrystalline materials

P18

Crystallization and preparation peculiarities of nanocrystalline powders based on metastable α′-tricalcium phosphate

Z. Zyman[1], M. Epple[2], D. Rokhmistrov[1], A. Goncharenko[1], O. Prymak[2], K. Loza[2]
[1]V.N. Karazin Kharkiv National University, Physics of Solids Department, Kharkiv, Ukraine
[2]University of Duisburg-Essen, Inorganic Chemistry and Centre for Nanointegration Duisburg-Essen (CeNIDE), Essen, Germany

Introduction

The purpose of this study was to develop a simple preparation route of metastable α-tricalcium phosphate (α′-TCP) nanopowders for effective but economical calcium phosphate cements (CPCs).

Materials and methods

The initial amorphous calcium phosphate (ACP) was prepared by the fast nitrate synthesis for known reagents with a Ca/P ratio of 1:1 at pH 10, 5 °C and a short precipitation time of 1 min [1]. The precipitate was washed, lyophilized (-23 °C, 24 h) and additionally dried in vacuum (48 h). The resulted dry powder was subjected to an abundant washing with an increased volume of distilled water (3 L, pH 10, 3 h), filtered and lyophilized (-20 °C, 24 h) [2]. The resulted powder was then heated to 1000 °C (heating rate 5 K/min and short annealing at a desired temperature), left to cool to RT and analyzed. XRD, TG/DTA, SEM, EDX and IR methods were employed for analysis.

Result and discussion

The lyophilized powder was XRD amorphous (an ACP) and crystallized at 630–700 °C into α′–TCP, which gradually transformed to stable β-TCP between 800 °C to 1000 °C. Consequently, in the ACP during abundant washing, besides usual rinsing the by-product of the synthesis, its hydrolysis (conversion) to an ACP" with higher Ca/P ratio (from initial 1.0 to 1.5) must have occurred. The formed α′ –TCP was slightly compacted and then ground to a dispersed nanocrystalline powder. This turned out to be an excellent component for a new CPC. The superior properties of the new α′–TCP (and the resulting CPC) are assumed to be associated with different nucleation clusters formed in hydrolyzed ACP′ compared to those in a traditionally processed one.

Conclussion

A simplified processing route was developed for the α′ –TCP nanocrystalline powder component in CPCs to make them both effective and convenient to apply.

References

[1] Zyman ZZ et al., J Mater Sci Mater Med, 21:123–130 2010
[2] Goncharenko A et al., ESB 2019, Abstracts, 893–894 2019

P19

The crystalline nature of zinc oxide nano- and microparticles with different shape determined by X-ray powder diffraction

O. Prymak[1], M. Olejnik[2], M. Epple[2]
[1]University of Duisburg-Essen, Inorganic Chemistry, Essen, Germany
[2]University of Duisburg-Essen, Inorganic Chemistry, Essen, Germany

Introduction

Zinc oxide particles are used in materials science and biomedicine where the size and anisotropic shape are decisive for their physicochemical and biological properties [1]. Therefore, an in-depth crystallographic understanding for the formation of monodisperse ZnO particles with a defined and uniform morphology is necessary to describe their crystal growth as determined by the reaction parameters [2].

Objectives

ZnO particles with well-defined size and shape were subjected to an extensive crystallographic investigation by X-ray powder diffraction, combined with Rietveld refinement. Lattice parameters, anisotropic crystallite size, and texture coefficients were determined.

Materials and methods

PVP-functionalized ZnO particles were synthesized by variation of the reaction parameters, purified and freeze dried and subsequently characterized by colloid-chemical, spectroscopic, thermogravimetric, microscopic, and crystallographic methods.

Results

The analyses confirmed a successful fabrication of monodisperse and uniform ZnO nano- and microparticles in five different shapes: Spheres 30 and 500 nm, rods 35×100 and 75×400 nm, plates 400×130 nm. Due to the preferred orientation of rods and plates on the sample holder, texture and crystal growth in [001] for rods and [h00] for plates with anisotropic crystallite size could be shown. In contrast, nano- and microspheres were isotropic and had a similar crystallite size, confirming their single-crystalline and polycrystalline nature, respectively.

Figure 1: Representative X-ray diffractogram (top) and scanning electron micrograph (bottom) of anisotropic ZnO nanorods with determined crystallite size (CrSize).

Conclusion

Rods and plates were anisotropic single crystals with different crystal growth directions whereas microspheres were isotropic polycrystals that consisted of smaller nanospheres.

References

[1] L. Cheng et al, *Nanoscale* 5 (2013) 3547
[2] L. Ann et al, *J. Environ. Chem. Eng.* 3 (2015) 436

Figure 1

hkl	[illegible]	002	101	102	110	103	200	112	201	004	202	104	203
CrSize / nm	35	70	39	45	39	51	35	42	40	71	40	53	40

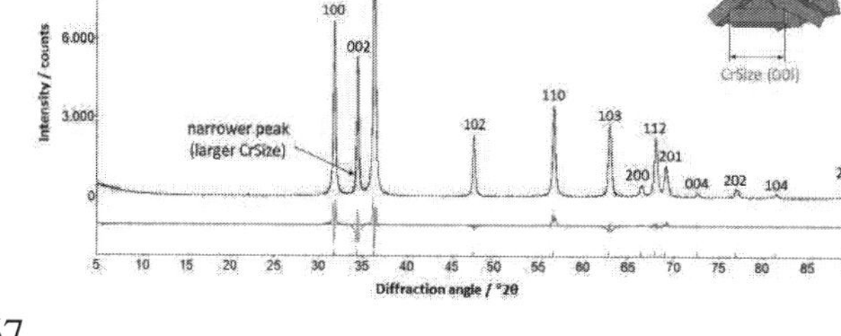

Figure 2

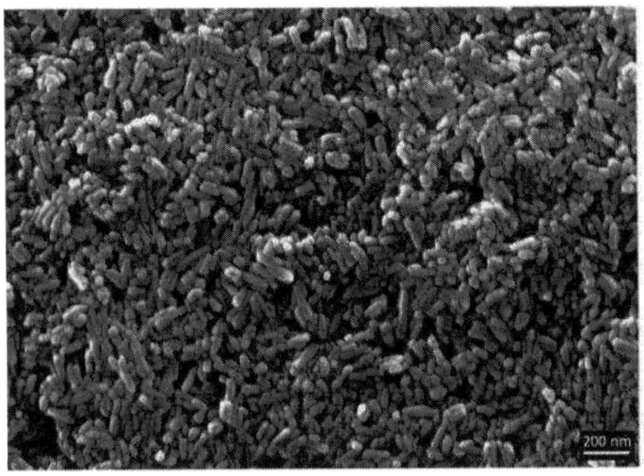

P20

Freezing of water droplets on glass surfaces with micro-scale lattice-patterned grooves

S. Yonezawa[1], T. Waku[1], Y. Hagiwara[1]
[1]Kyoto Institute of Technology, Kyoto, Japan

The freezing of water droplets deposited on surfaces causes serious troubles, such as low visibility of traffic lights, traffic mirrors and surveillance cameras. The droplets in these troubles are often in a supercooled state. However, the experimental results for the freezing of supercooling droplets on cold surfaces are limited. Thus, we carried out measurements on the freezing of deposited cooled droplets (or supercooled droplets) on glass surfaces. We adopted glass plates with micro-scale lattice-patterned grooves because the grooves had attenuated the growth of frost crystals [1], which can be a trigger of the droplet freezing. In addition, we coated the glass surfaces with a silane coupling agent to reduce the contact area of droplets. We analyzed the images of deposited droplets captured with a high-speed video camera. The measurement results showed that the micro-scale lattice-patterned grooves reduced the apparent contact areas of deposited droplets. The grooves increased the completion time of droplet freezing (See Fig. 1). In addition, the completion time of ice-shell formation was prolonged by the surface coating with the coupling agent. Thus, it is concluded that the newly fabricated surfaces in this study are promising.

[1] S. Yonezawa, H. Agui, H. Ohkubo and Y. Hagiwara, Suppression of frost layer growth on glass or copper surfaces with micro-scale lattice-patterned grooves, Int. J. of Refrigeration, vol. 106 (2019). pp. 33–40.

Figure 1

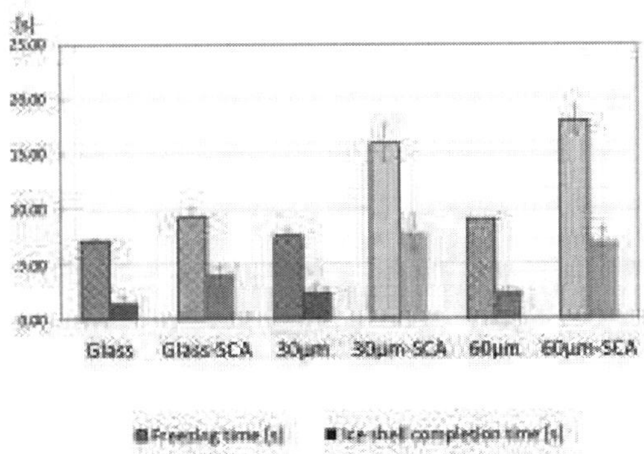

P21

Depth Profiling of Polycrystalline Cu(In,Ga)(Se,S)₂ Thin Films with Asymmetric Bragg Diffraction

J. Dallmann[1], M. Schuster[2], A. Weber[3], P. Wellmann[2], R. Hock[1]
[1]Friedrich-Alexander-University Erlangen-Nürnberg, Institute for Crystallography and Structural Physics, Erlangen, Germany
[2]Friedrich-Alexander-University Erlangen-Nürnberg, Institute for Materials for Electronics and Energy Technology, Erlangen, Germany
[3]AVANCIS GmbH, Munich, Germany

Polycrystalline $Cu(In_xGa_{1-x})(Se_yS_{1-y})_2$ thin films are well-established in photovoltaic applications. They have a composition dependent bandgap, allowing to tune the absorption spectra by a bandgap gradient throughout the thin films.

Hence, the reliable determination of the compositional depth profile is important for the characterization of chalcopyrite solar cells and for the optimization of their conversion efficiency [1].

Asymmetric Bragg diffraction provides the possibility to measure depth sensitive diffraction patterns by variation of the incidence angle [2]. Whole pattern Pawley fits allow precise determination of the lattice parameters of the phases present in the absorber. The composition of the chalcopyrite(s) at a certain penetration depth can be calculated from the lattice parameters. A few difficulties must be taken into account when measuring with an asymmetric setup at incidence angles as small or smaller than the critical angle, such as the reflection shift based on refraction. Another difficulty is that the signal of any particular depth cannot easily be isolated from the signal originating from layers closer to the surface.

The results from measurements from the top of the thin film with various penetration depths are compared to measurements from its backside. Furthermore, the thin films are chemically etched to various depths and measured without the influence of higher layers to complement and confirm the obtained depth profiles.

[1] W. Witte et al., *Prog. Photovolt: Res. Appl.* (2015). 23, 717-733
[2] I. M. Kötschau, H. W. Schock, *J. Appl. Cryst.* (2006). 39, 683-696

P22

Numerical simulation of the freezing impinged water droplets on a cooling surface

D. Utsumi[1], S. Fujii[1], Y. Hagiwara[1]
[1]Kyoto Institute of Technology, Kyoto, Japan

Introduction
The freezing water droplets cause many troubles in automobiles and heat exchanger. Thus, the suppression of the freezing of water droplets has been required. Several attempts have been carried out to elucidate the freezing of sessile droplets [1].

Objectives
However, few attempts of numerical simulation have been made to predict the freezing of impinged water droplets.

Methods
In this study, we carried out two-dimensional simulation on the freezing impinged water droplets on a cold surface using a Phase-field method. We adopted a phase-field method to identify the air/water interface and ice/water interface. We added a term to the

modified Allen-Cahn equation to predict a projection of frozen droplet. Also, the Navier-Stokes equation and energy equation were solved. An ice nucleus was allocated on a place where the temperature was lower than -15 °C. This is because water was supercooled before freezing [1].

Results and conclusion

The computational results showed that a projection was formed on the cap of a frozen droplet. The results also showed that an ice shell was grown along the droplet's surface (See Fig.1. The white, gray and black areas show water, air and ice, respectively.). The volume of ice was 8.1% higher than that of water droplet. In addition, the change in the mass due to freezing was only 0.82%. A concave shape of the freezing front was predicted, and the temperature field outside the droplet was reasonable. Thus, our computational method was verified.

Reference

[1] Y. Hagiwara, S. Ishikawa, R. Kimura, K. Toyohara, Ice growth and interface oscillation of water droplets impinged on a cooling surface, J. Crystal Growth vol468 (2017), pp.46-53.

Figure 1

P23

The sacrificial anode effect in a physical mixture of silver and platinum nanoparticles

K. Loza[1], M. Breisch[2], A. Rostek[1], O. Wetzel[1], M. Heggen[3], M. Köller[2], C. Sengstock[2], M. Epple[1]
[1]University of Duisburg-Essen, Inorganic Chemistry, Essen, Germany
[2]BG University Hospital Bergmannsheil Bochum, Bochum, Germany
[3]Research Center Jülich GmbH, Ernst Ruska-Centre (ER-C) for Microscopy and Spectroscopy with Electrons, Jülich, Germany

Silver is a well-known antimicrobial agent against a broad range of bacteria, viruses, and fungi [1]. Therefore, it is widely used in medical implants and instruments as well as textiles and cosmetics [2]. The biological effect of Ag nanoparticles is based on the oxidative release of silver ions that interact with cellular structures such as cell wall components, enzymes, or DNA, inducing adverse biological effects [3]. The antimicrobial efficiency can be increased by surface enhancement (nanoparticles) or by creating a sacrificial anode system for Ag. Here, the enlarged surface of nanoparticles with a presence of the electrochemically more noble platinum in physical mixtures of Ag and Pt nanoparticle dispersions were combined.

The internal structure and the elemental distribution within the particles in different saline media were elucidated by scanning transmission electron microscopy in combination with energy-dispersive X-ray spectroscopy. The results revealed that multiple etching contributed to the morphology evolution during silver nanoparticle corrosion. The preferential corrosion sites were corners. Silver atoms (having a lower reduction potential) oxidize and dissolve into solution, thus sacrificing the nanoparticle. Galvanic exchange reactions led to redeposition of the metal and an alloying process.

It was found that the dissolution rate of silver nanoparticles was enhanced in the presence of platinum nanoparticles within the physical mixture compared to a dispersion of pure silver nanoparticles. Dissolution experiments revealed a fourfold increased Ag ion release from physical mixtures due to enhanced electrochemical activity, which resulted in a significantly increased toxicity towards both bacteria and human mesenchymal stem cells.

References

[1] A. Panacek et al., Biomaterials, 2009, 30, 6333–6340.
[2] M. Epple et al., Angew. Chem., Int. Ed., 2013, 52, 1636–1653.
[3] M. Koeller et al., Mater. Sci. Eng., C, 2017, 74, 536–541.

P24

How brachiopods generate a hybrid composite shell material by cellular control of crystal growth of calcite - a biomineralogic SEM/TEM study

E. Griesshaber[1], A. Ziegler[2], M. D. M. Simonet Roda[1], X. Yin[1], U. Rupp[2], P. Walther[2], D. Henkel[3], A. Checa[4], V. Häussermann[5], W. Schmahl[1]
[1]Ludwig-Maximilians-University Munich, Department for Earth- and Environmental Sciences, Munich, Germany
[2]Universität Ulm, Central Facility of Electron Microscopy, Ulm, Germany
[3]GEOMAR Helmholtz Centre for Ocean Research , Marine Biogeochemistry/Marine Systems, Kiel, Germany
[4]University of Granada, Stratigraphy and Paleontology (Faculty of Sciences), Granada, Spain
[5]Huinay Scientific Field Station, Puerto Montt, Chile

Brachiopods are marine invertebrates forming fracture-toughened shells. Their shell is a hybrid composite of calcite (CaCO3) mineral fibers in ca. 2 w.% biopolymer matrix. The shell is structured into a thin, outer, organic periostracum, followed inwardly by a thin, hard, nanostructured calcite layer, which coats the main, ductile, fiber composite layer. Each calcite fiber is a single-crystal with two concave sides and a convex side which is lined by a biopolymer membrane (Fig.1). The fibers are assembled into a characteristic space-filling pattern, where the convex side interlocks with concave sides of adjacent fibers (Fig. 1). The pattern emerges from the close control of crystal growth by the mineralizing outer mantle epithelial (OME) cells. Ca2+, CO2/HCO3- and H+ transport occurs through the cell membrane, which is in tight contact with the convex side of the growing crystal. Here, crystal growth occurs in a concentric manner until terminated when the cells form an extracellular organic membrane lining the convex growth surface. The mantle epithelium cells connect to this membrane lining by apical hemidesmosomes, thus attaching the mantle epithelium to the shell.

Our SEM and TEM images of large shell/cells interface regions suggest a sequence of coordinated actions as follows: (i) local detachment of epithelial cell membrane from the extracellular membrane of previously formed fibers; (ii) CaCO3 mineralization above these detachments; (iii) after the final width of the fiber is reached, secretion of an extracellular membrane starting from the edges of the fiber and progressing to the center of the convex fiber surface; mineral secretion is suspended where this extracellular membrane is formed; (iv) attachment of the cells by apical hemidesmosomes to the newly formed extracellular organic membrane.

Simonet Roda, M. et al. (2019) Terebratulide brachiopod shell biomineralization by mantle epithelial cells. *J. Struct. Biol.* 207: 136-157

Fig. 1 See text

Figure 1

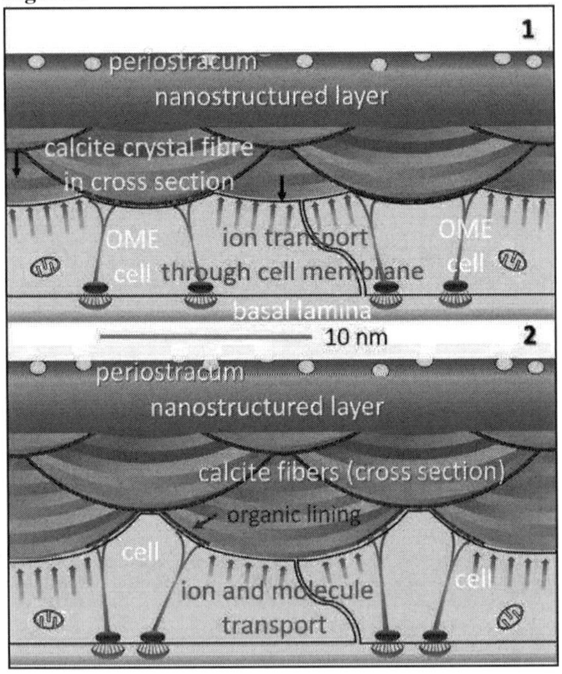

P25

Improved algorithm to calculate the powder pair distribution function PDF

R. Neder[1]
[1]Friedrich-Alexander-University Erlangen-Nürnberg, Institute for Crystallography and Structural Physics, Erlangen, Germany

The usual PDF algorithms as in DISCUS of PDFgui calculate the PDF in direct space. The histogram is convoluted by a thermal distribution function and converted by multiplying with a quotient of pair wise atomic form factors divided by the average atomic form factor. The implicit assumption is that the quotient of the form factors is constant over the entire Q range. The deviation of this quotient from a constant value causes intensity variations in the calculated PDF as will be shown for a few examples. As these deviations are of the same order of magnitude as the differences between experimental and model PDF s in modern simulations, this systematic error cannot be tolerated any longer.

The new algorithm for the calculation of the PDF that is based on an accurate calculation of the normalized total scattering function S(Q). The PDF is calculated by the sine Fourier transform.

As the calculation of the powder pattern is accurate within the limits of the kinematic approximation, this new algorithm provides an accurate PDF for neutron, X-ray and electron diffraction. The advantages of the new algorithm are

Exact: The function is accurate for x-ray and electron diffraction

Computational speed: The algorithm does not depend on the maximum distance R

Initial date check: Errors like zero point shifts, wave length errors are easily detected

Preferred Orientation: Algorithms from Rietveld calculations allow the simulation of preferred orientation effects, which are difficult in direct space calculations

Magnetic scattering: Can be calculated straightforwardly in a powder pattern

Instrumental resolution: Arbitrarily complex resolution functions can be applied as a convolution in reciprocal space.

Sample contribution: For large models, strain can be described as a convolution of the powder pattern

Nanoparticles: By using the Debye-Scattering-Equation, arbitrarily shaped nanoparticles will result in the accurate PDF without artificial corrections of the baseline

P26

Characterizing phase generation and structural effects due to micronization and amorphization of an active pharmaceutical ingredient for HIV treatment

M. Terban[1], L. Russo[2], T. Pham[2], D. Barich[3], Y. Sun[3], M. Burke[4], J. Brum[3], S. Billinge[5,6]
[1]Max-Planck-Institute for Solid State Research, Scientific Facility X-ray diffraction, Stuttgart, Germany
[2]GlaxoSmithKline, R&D, Stevenage, United Kingdom
[3]GlaxoSmithKline, R&D, Collegeville, United States
[4]GlaxoSmithKline, R&D, King of Prussia, United States
[5]Columbia University, Applied Physics and Applied Mathematics, New York, United States
[6]Brookhaven National Laboratory, Condensed Matter Physics and Materials Science, Upton, United States

Processing procedures for controlling crystallite size and the amorphous phase content are critical for formulating drug products with the desired bioavailability and product stability. To improve success rates during early drug development stages, an understanding of disordered phase generation and structural changes is very important. An example is an active pharmaceutical ingredient (API) currently being developed for the treatment of the human immunodeficiency virus (HIV) by inhibiting the enzymatic activity that propagates infectious activity. The API (GSK2838232B) is practically insoluble in gastric and intestinal fluids, but the in vivo activity can be improved through particle size reduction.

In this study, the efficacy of distinguishing and quantifying different structural states by total scattering and pair distribution function (PDF) analysis was assessed [1,2]. Varying degrees of crystallite size reduction and amorphous phase generation were achieved through different micronization processes including jet air milling, ball milling, hand grinding, and microfluidics processing. Structural refinements were used to characterize the crystalline, micronized, and amorphous formulations, and the PDF data were benchmarked for quantifying amorphous phase generation in the API formulations. Solid state nuclear magnetic resonance (SSNMR) was performed to complement the structural assessment and phase quantification.

Disordering mechanisms were found to vary for different processing procedures, and differences in both local structure and dynamics were observed for differently prepared amorphous formulations. PDF and SSNMR were found to agree well for the quantification of amorphous phase in different formulations.

[1] Terban, M. W.; Cheung, E. Y.; Krolikowski, P.; Billinge, S. J. L. Cryst. Growth Des. 2015, 16, 210220.
[2] Morrow, E. A.; Terban, M. W.; Lee, J. W.; Thomas, L. C.; Billinge, S. J. L.; Schmidt, S. J. J. Food Eng. 2019, 261, 87–99.

P28

Synthesis of bimetallic Nickel nanoparticles as catalysts for the Sabatier reaction

M. Heilmann[1], R. Bienert[1], C. Prinz[1], F. Emmerling[1]
[1]Federal Institute for Materials Research and Testing (BAM) , Structural Analysis, Berlin, Germany

Nanoparticles (NP) have become important materials for a variety of chemical technologies. The enhance surface-area-to-volume ratio of NPs is very high, making them excellent for use as catalyst, in analytical assays, and for antimicrobial applications. [1-3]

Nickel NPs have exhibited immense potential as important catalyst for the Sabatier reaction, *i.e.* converting waste to energy *via* transformation of CO2 into CH4,[4] and could replace the rare earth elements such as Ru, PT, or Rh. In this work we describe the solvothermal synthesis of monometallic and bimetallic nickel nanoparticles.

Monodisperse monometallic Ni nanoparticles were synthesized using Oleylamin as solvent and reducing agent. The nanoparticles were investigated using small angle scattering (SAXS), scanning transmission electron microscopy (STEM) and energy dispersive X-ray spectroscopy (EDX), showing that the catalytically active sites are accessible. However, Ni has a high propensity to undergo oxidation, and becoming deactivated by coke formation.[5] Hence, we further explore the preparation of bimetallic NPs, where a second metal is added to stabilize the Ni.[6]

Bimetallic Cu-Ni NPs were synthesized by simultaneous solvothermal reduction. These bimetallic NPs exhibit excellent catalytic properties are promising candidates to be used as catalysts for efficient energy storage.

References

[1] S. Kawi, Y. Kathiraser, J. Ni, U. Oemar, Z. W. Li, E. T. Saw, *Chemsuschem* 2015, *8*, 3556-3575.
[2] T. S. Bedwell, M. J. Whitcombe, *Anal. Bioanal. Chem.* 2016, *408*, 1735-1751.
[3] S. P. Deshmukh, S. M. Patil, S. B. Mullani, S. D. Delekar, *Mater. Sci. Eng. C-Mater. Biol. Appl.* 2019, *97*, 954-965.
[4] A. Maroufmashat, M. Fowler, *Energies* 2017, *10*, 22.
[5] C.-j. Liu, J. Ye, J. Jiang, Y. Pan, *ChemCatChem* 2011, *3*, 529-541.
[6] Wu, H. J.; La Parola, V.; Pantaleo, G.; Puleo, F.; Venezia, A. M.; Liotta, L. F. *Catalysts* 2013, *3*, 563-583.

P29

Observations of Crystalline Microstructure by Exploring Powder X-Ray Diffraction Patterns

M. Kojdecki[1], P. Pardo[2], J. M. Calatayud[2], J. M. Amigó[2], J. Alarcón[2]
[1]Military University of Technology, Institute of Mathematics and Cryptology, Warsaw, Poland
[2]University of Valencia, Department of Inorganic Chemistry, Valencia, Spain

X-ray diffraction is a principal phenomenon exploited to characterising the crystalline microstructure of nanomaterials. An adequate mathematical model of a material, an accurate description of a corresponding X-ray diffraction pattern and a stable method for solving an ill-posed inverse problem involved are necessary for determining microstructural parameters of a polycrystal reliably from experimental X-ray diffraction data. An effective procedure for determining microstructural characteristics of polycrystalline materials has been developed recently in subsequent works, based on ideas proposed earlier. The computational method exploited simultaneously several line profiles extracted from powder X-ray diffraction pattern for a specimen. Quantitative characteristics like a prevalent crystallite shape, a volume-weighted crystallite size distribution and a second-order crystalline lattice strain distribution were computed for several polycrystalline materials. These microstructural characteristics enabled unique insight into formation of polycrystalline phases in processes of crystallite nucleation and growth. In the presentation the procedure will be concisely described and an example of the microstructure of boehmite will be presented and discussed briefly. The boehmite nanoparticles were prepared in a procedure comprising the formation of a precipitate and subsequent hydrothermal aging with possible application of an additive. Based on the computed characteristics, three types of the microstructure correlated with the production procedures were observed. The nanoparticles of boehmites were found to be monocrystalline grains with characteristic habits and sizes of order of ten nanometers and with bimodal logarithmic-normal size distributions.

P Pardo, M A Kojdecki, J M Calatayud, J.M. Amigó, J. Alarcón: Crystalline microstructure of boehmites studied by multi-peak analysis of powder X-ray diffraction patterns; Powder Diffraction, 32 (S1) (2017), S87-S98.

Solid state physics in crystallography

P31

Synthesis of Elasto-Plastic Deformable Ceramics by High Pressure

M. Araki[1]
[1]Starship Res. Inst., Handa, Japan

Ceramics were sintered with metals under very high pressure of 1 GPa and temperatures over 2,000 C. After the sintering, metals with single phase did not remain as metal and resulted materials consisted of metal rich ceramics with unkown phases. Sintered ceramics were tested and measured by compressions. They exhibited initially elastic and later plastic deformations. Some of them had shown irregular deformations. The reasons of the irregular deformations may have happened by voids formed while the high pressure- high temperatuure operations. For avoiding the voids, raising the pressure level should be effective. Table 1. showing some of test conditions and result, Fig. 1 shows sintered ceramics and Fig. 2 showing one of deformation-stress compression test result.

(1) This work was achieved under support by Chubu Denryoku K.K. (Central Electric Power Co. Ltd.)

(2) Physical properties were measured by Quantum Physics Lab. and Biological System Lab. of Mie University.

(3) Originally reported at 48th Koatu Touron Kai (Special Issure of High Pressure Science and Technology Nov. 20 2007 Kurayosi Japan)

Figure 1

Table 1. Compositions, Sintering Temperatures and Results

No.	Compositions Vol.%	Sintered Temp. C	Result of X ray Analisys (Number of Peaks)
1	TiC - Ti 24.6 - Mo 17.7	2,600	TiC(3), MoC(3), Unkown(18)
2	TiC-Ti4.4-Al19.1-Mo12.6	2.220	TiC(3), MoC(4), α Mo_2C(1), unkown(11)
3	TiC-Ti16.9-Nb12.2-Mo12.7	3,140	TiC(3), Nb_xC_y(3), NbC(1), MoC(3), unkown(11)

Figure 2

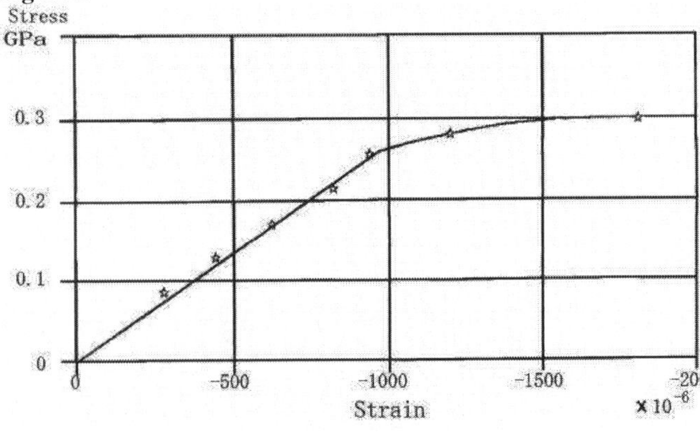

Fig.1 Compression Test result Specimen Type No.1

Run No. CDR 31

P32

Primary crystal orientation of thin-walled area of single-crystalline turbine blade airfoils

J. Krawczyk[1], W. Bogdanowicz[1], R. Paszkowski[1]
[1]University of Silesia in Katowice, Institute of Materials Engineering, Chorzów, Poland

The single-crystalline superalloys are a group of materials characterized by increased strength properties, especially the creep resistance at high operating temperatures. Crystal orientation, which depends on the parameters of directional crystallization and cast geometry, have an important effect on mechanical properties of the element. In thin-walled areas the volume fraction of the regions where the residual stress occurs, is higher than in the thick blade parts, that contributes to local changes in crystal orientation.

The aim of the study was to determine a distribution of the α angle describing the primary crystal orientation of analysed blade airfoils, and to analyse the relation between this distribution and the slope of mould walls in relation to the axis of blades obtained at different withdrawal rates. An additional aim was to consider the mould walls inclination on the growth direction of the primary dendrite arms.

The thin-walled airfoil areas of as-cast single-crystalline turbine blades made of CMSX-4 superalloy were studied. The blades were produced by industrial Bridgman technique at the withdrawal rates of 2, 3 and 4 mm/min. The angle between the [001] crystallographic direction and blade axis, related to the primary orientation, was defined by Ω-scan X-ray diffraction method and by the Laue diffraction at selected points of the camber line. The effect of change in the [001] crystallographic direction was considered with regard to the arrangement of primary dendrite arms in relation to the trailing edge and the camber line. It was stated that when the distance between the mould walls is less than the critical value of about 1.5 mm. The directing effect increases as the distance between the walls of the mould decreases. The effect may be controlled by select an appropriate secondary orientation using seed crystal in the blade production process. The model of dendrite interaction with the mould walls including bending and "deflection" was proposed.

P33

Crystal structure and pseudo-symmetry of the 3D-weak topological insulator Bi12Rh3Cu2I5.

E. Carrillo[1], R. Ray[2], M. Richter[2,3], M. Ruck[4,1]
[1]Technische Universität Dresden, Faculty of Chemistry and Food Chemistry, Dresden, Germany
[2]IFW Dresden, Dresden, Germany
[3]Technische Universität Dresden, Dresden Center for Computational Materials Science, Dresden, Germany
[4]Max-Planck-Institut für Chemische Physik fester Stoffe, Dresden, Germany

Introduction

A series of bismuth subiodides: $Bi_{14}Rh_3I_9$, $Bi_{13}Pt_3I_7$, $Bi_{12}Pt_3I_5$, showing stacks of quantum spin hall (QSH) layers $[(Bi_{8/2}M)_3I]^{X+}$ (M = Pt, Rh) have been studied [1, 2]. The appearance of the non-trivial topological properties is highly regulated by their stacking sequence and type of spacer. Thus, $Bi_{14}Rh_3I_9$ is a 3D-weak topological insulator, while $Bi_{13}Pt_3I_7$ is a semimetal due to the "dimerization" of QSH layers allowed by a one atom thick spacer composed only by iodine [2]. New compounds are searched for understanding and tuning of the properties of this class of materials.

Materials and methods

Bi12Rh3Cu2I5 was synthesized slow cooling a melt of Bi, Rh, Cu and BiI3, following a temperature program based on thermochemical data.

Results

X-ray diffraction on a single-crystal of $Bi_{12}Rh_3Cu_2I_5$ revealed a monoclinic lattice (space group *C*12/*m*1). The QSH layer $[Bi_{12}Rh_3I]^{2+}$, which determinates the electronic states with inverted parity around the Fermi level, is a Kagome net of edge-sharing $[RhBi_{8/2}]$ cubes with iodide ions in its hexagonal prismatic voids. The anionic spacer layer is composed of discrete $[Cu_2I_4]^{2-}$ units, tilted 7.02(2) ° out of the (002) plane. The layered nature of the compound produces several faults along the stacking direction. Although the spacer layer is only one atom thick, full-relativistic DFT-based calculations show a topological gap of ≈10 meV. The cell metrics complies with an *F*-centered pseudo-orthorhombic (oF) lattice, which itself has an pseudo-orthohexagonal basis ($b_{oF}/a_{oF} = 1.72 \approx \sqrt{3}$). This higher metric symmetry causes pseudo-merohedral twinning: the studied crystal shows a mirror-twin in the oF (001) plane, with fractional contributions of about 0.89:0.11. The lateral offset of neighboring layers of the same kind is exactly *a*/2.

Figure: Crystal structure of Bi12Rh3Cu2I5, also showing the pseudo-oF cell edges.

[1] Rasche, B. et al. *Nature Materials.* 12 (2013): 422.
[2] Pauly, C. et al. *Nature Physics.* 11 (2015): 338.

Figure 1

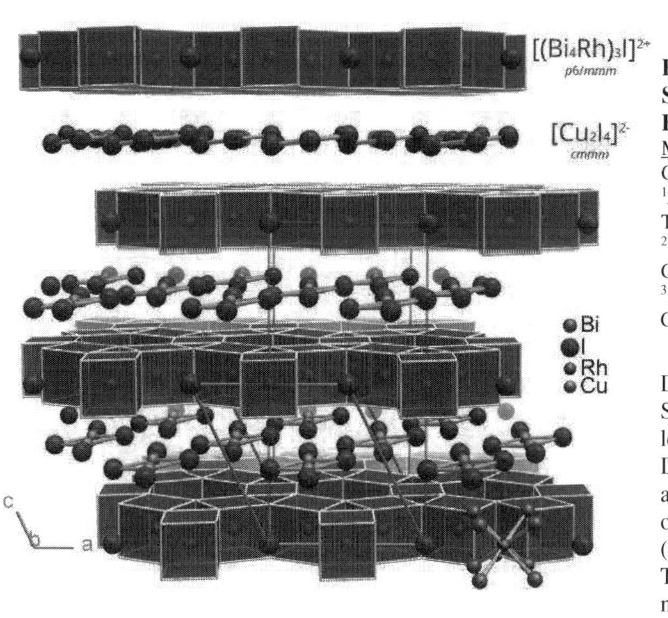

P34

Multi-scale defect analysis in single-crystalline nickel-based superalloys

R. Paszkowski[1], J. Krawczyk[1], W. Bogdanowicz[1]
[1]University of Silesia in Katowice, Institute of Materials Engineering, Chorzów, Poland

Nickel-based superalloys due to unique strength properties are used for production of many structural elements, inter alia, the high pressure turbines blades in aircraft engines [1]. Due to the specific structure of superalloys and complex shape of the blades, various growth defects may be formed during crystallization process, that are not eliminated in subsequent stages of production. Even a small number of defects arising during crystallization may cause a decrease in strength properties. For this reason, it is important to analyse the as-cast structure of the blades.

The as-cast single-crystalline turbine blades made of industrial nickel-based superalloy of CMSX-6 were studied. They were obtained by directional solidification of Bridgman technique using an industrial ALD Vacuum Technologies vertical furnace [2]. Analysis of crystal microstructure was performed using the Scanning Electron Microscopy, X-ray diffraction topography, X-ray diffraction mapping of primary crystal orientation and lattice parameter $a_0^{\gamma''}$. In addition, using the positron annihilation lifetime spectroscopy (PALS), the distribution of defects in areas determined by X-ray topography was analyzed, which allowed to determine the presence of vacancies and clusters of vacancies [3].

Based on obtained results it was found that the simultaneous use of scanning electron microscopy, X-ray methods and positron annihilation lifetime spectroscopy allows to multi-scale analysis of the type and distribution of defects in the area of the root directly adjacent to the selector, from which the defects are inherited to other parts of the blade.

Refererences

[1] R.C. Reed, The Superalloys, Fundamentals and Applications, Cambridge University Press, Cambridge 2006.
[2] A. Onyszko, K. Kubiak, J. Sieniawski, Journal of Achievements in Materials and Manufacturing Engineering, 32, 66 (2009)
[3] J. Krawczyk, R. Paszkowski, W. Bogdanowicz, A. Hanc-Kuczkowska, J. Sieniawski, B. Terlecki, Materials 12, 870 (2019)

P35

Solid-state characterization and solubility enhancement of BCS-class II drug-dasatinib

M. S. T. Abdul Azeeze[1], P. Palanisamy[2], M. Subramania Nainar [1], C. Ghoroi[3]
[1]JSS College of Pharmacy-Udhagamandalam, Pharmaceutical Analysis, The Nilgiris, India
[2]Indian Institute of Technology Gandhi Nagar, Biological Engineering, Gandhi Nagar, India
[3]Indian Institute of Technology Gandhi Nagar, Chemical Engineering, Gandhi Nagar, India

Dasatinib (DAS) is a BCS (Biopharmaceutical Classification System) class II drug used for the treatment of chronic myeloid leukaemia. The main aim of this study is to improve the solubility DAS. The DAS-4-CBA (4-chloro benzoic acid) salt was prepared at room temperature (25°C) by the solution crystallization method. The salt was characterized by Single Crystal X-ray diffraction (SCXRD), followed through the salt was subjected to Fourier Transform-Infra red Spectroscopy (FT-IR) to find out the intermolecular interactions about the formation of salt. The results shows that the acid proton is transferred to the secondary N atom on the piperazine ring of DAS, forming the charged -assisted hydrogen bond $N\text{–}H^{+}\cdots O^{-}$. The refined structure represented that asymmetric unit of the DAS-4-CBA salt comprises one DAS, one 4-CBA and two DMF molecules. Molecular packing of DAS-4-CBA salt is facilitated through a hydroxyl group of DAS and the carbonyl oxygen of 4-CBA. The thermal stability of the grown crystal was analyzed using differential scanning calorimetry (DSC). Solubility studies of DAS and DAS-4-CBA were conducted and they exhibited 7.00 fold increased solubility respectively when compared to that of the pure drug.

Keywords: Dasatinib, DSC, FT-IR, Solubility, Salt, SCXRD

Fig.1 Graph set patterns D1, 1(2) a, b, c and d of DAS-4-CBA held together by N-H···O and O-H···O hydrogen bonds

Fig.2 2D-layer of DAS-4CBA held together by N-H···O and O-H···O hydrogen bonds

References
[1] Dolomanov O.V, Bourhis L.J, Gildea R.J, Howard J.A.K, and Puschmann H. (2009).OLEX.J. Appl Cryst.42: 229–341.
[2] http://www.olexsys.org/
[3] Food and Drug Administration, Washington (USA): Dasatinib (Sprycel), Application number: 21-986 & 22-072, Clinical Pharmacology and Biopharmaceutics Review(s).
[4] Desiraju, G. R.: "Crystal Engineering. The Design of Organic Solids", Elsevier, 1989.

Figure 1

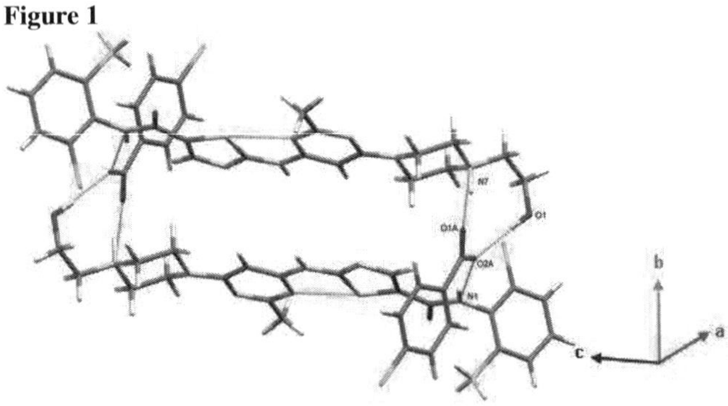

Figure 2

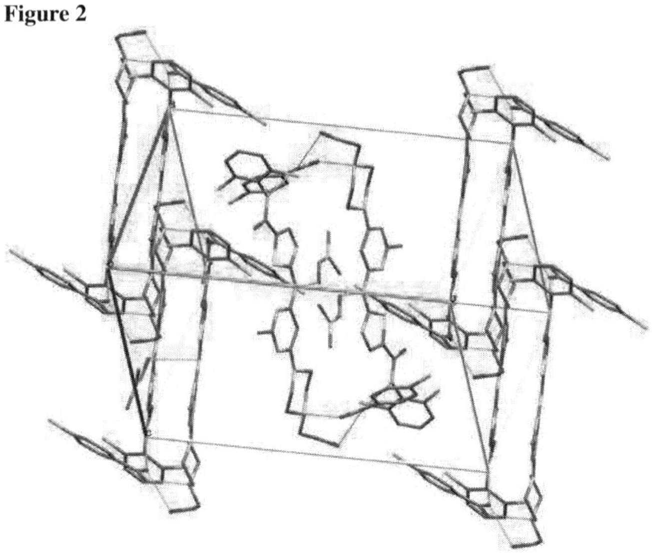

P36

Temperature-dependent EXAFS measurements of the lead L3 edge allowing quantification of the anharmonicity of chlorine-substituted methylammonium (MA) lead triiodide

G. Schuck[1], D. M. Többens[1], D. Wallacher[1], N. Grimm[1], S. Schorr[1,2]
[1]Helmholtz-Zentrum Berlin für Materialien und Energie GmbH, Berlin, Germany
[2]Freie Universität Berlin, Institute of Geological Sciences, Berlin, Germany

Research interest has increasingly focused on hybrid perovskites $MABX_3$ like $[CH_3NH_3]^+$ (MA), B = Pb and X = I or Cl as a future photovoltaic material. The analysis of the EXAFS Debye-Waller factor of chlorine-substituted methylammonium (MA) lead triiodide allows a direct determination of the influence of chlorine substitution on the anharmonicity of the lead-halid bond. In the ordered orthorhombic crystal structure, chlorine substitution leads to a reduction of the anharmonic fraction (cumulant C_3) of the radial pair distribution function. Here, the behavior of C_3 can be described in a classical approximation by the force constants k_0 (harmonic fraction) and k_3 (anharmonic fraction). While k_0 for the three investigated halides does not show any major differences ($MAPbI_3$ k_0 = 1.11(2) mdyn/Å; $MAPbI_{2.94}Cl_{0.06}$ k_0 = 1.12(1) mdyn/Å; $MAPbCl_3$ k0 = 1.23(3) mdyn/Å), there are clear differences for k_3 ($MAPbI_3$ k_3 = 0.696(6) mdyn/Å^2; $MAPbI_{2.94}Cl_{0.06}$ k_3 = 0.665(4) mdyn/Å^2; $MAPbCl_3$ k_3 = 0.46(3) mdyn/Å^2). Also, the perpendicular part of the EXAFS Debye-Waller factor $\sigma^{\perp}$, whose temperature-dependence can be described by the Einstein temperature E ($MAPbI_3$ E = 38.9(5) K; $MAPbI_{2.94}Cl_{0.06}$ E = 43.3(6) K; $MAPbCl_3$ E = 94(2) K) shows that chlorine substitution in $MAPbI_3$ in the orthorhombic phase leads to a reduction of the anharmonicity. FTIR [1] and quasi-elastic neutron scattering (QENS) [2] studies showed that chlorine substitution has a large influence on the rotational dynamics of the MA molecule in $MAPbI_{3-x}Cl_x$ perovskites[1] since the chlorine substitution leads to a weakening of the hydrogen bridge bonds (these bonds connect the MA molecules with the [PbX6]- octahedra host structure).[2] Here, we discuss the results of the cumulant analysis in context of the MA molecule rotational dynamics changes due to chlorine substitution, which we already investigated with QENS.

[1] G. Schuck, et. al., J. Phys. Chem. C, 2018, 122, 5227
[2] G. Schuck, et. al., J. Phys. Chem. C, 2019, 123, 11436

Bio-Crystallography I: Signalling, macromolecular interactions and other new structures

P37

Structural and biochemical studies of human mitochondrial post-transcriptional regulator FASTK

D. Dawidziak[1,2], M. Kuska[1], J. Kutner[1], M. Merski[3], M. Górna[1,3], K. Woźniak[1]
[1]University of Warsaw, Core Facility for Crystallography and Biophysics, Biological and Chemical Research Center, Department of Chemistry, Warsaw, Poland
[2]University of Warsaw, College of Inter-Faculty Individual Studies in Mathematics and Natural Sciences, Warsaw, Poland
[3]University of Warsaw, Structural Biology Group, Biological and Chemical Research Center, Department of Chemistry, Warsaw, Poland

The main scientific goal of the project is to study the function and structure of the human mitochondrial post-transcriptional regulator FASTK (Fas-activated serine/threonine kinase). This molecule is involved in the regulation of gene expression in mitochondria at the level of RNA processing. Moreover, FASTK was confirmed to bind RNA, but little is known about its three-dimensional structure and RNA specificity.

We report first successful protocols for obtaining soluble constructs of human FASTK as well as first promising crystallization attempts which should enable its further biochemical and structural characterization. By employing RNA-binding and activity assays like EMSA we aim to determine the model of its cooperation with the mitochondrial degradosome and its role in the regulation of ND6 mRNA biogenesis. Our recent EMSA results confirmed binding of FASTK construct to short fragments of mitochondrial ND6 mRNA what should allow us co-crystallization of FASTK-RNA complexes. Molecule was also prepared for SEC-SAXS measurements.

FASTK has potentially new RNA binding structural domains: an N terminal helical domain of unknown structure and a potentially unusual PD(D/E)XK nuclease domain. Due to the involvement of FASTK protein in pro-inflammatory processes, obtaining a structural model could allow to design anti inflammatory drugs to treat autoimmune diseases such as asthma or rheumatoid arthritis.

This work was supported by Foundation for Polish Science/European Union under the European Regional Development Fund (TEAM TECH CORE FACILITY/2017-3/4, POIR.04.04.00-00-31DF/17-00) and by National Science Centre, Poland, under the grant agreement number 2014/15/D/NZ1/00968
.

P38

Core Facility for Crystallographic and Biophysical Research to support the development of medicinal products

J. Kutner[1], M. Gorna[1], M. Malińska[1], M. Wanat[1], D. Dawidziak[1], K. Polak[1], M. Kuska[1], W. Lidwin[1], M. Kisiala[1], S. Sutula[1], K. Wozniak[1]
[1]University of Warsaw, Biological and Chemical Research Centre, Department of Chemistry, Warsaw, Poland

As a result of TEAM-TECH Core Facility Project from the Foundation for Polish Science, we have established the Core Facility for Crystallography and Biophysics (CFCB) at the Biological and Chemical Research Centre, University of Warsaw, under the supervision of **Professor Krzysztof Woźniak** (Head) and **Jan Kutner, Ph.D.** (Deputy Manager).

The mission of the Core Facility is focused on the analysis of proteins and small molecule compounds leading to crystallization trials for academic and commercial users. The project enables studies of challenging biochemical and pharmaceutical problems, with emphasis on drug development. Research at CFCB is carried out in an interdisciplinary way, including both wet biology (**"BIO"**) and chemical crystallography (**"CHEM"**) techniques as well as theoretical approaches including structure modelling, bioinformatics and computational methods. Biology and chemistry team members work in synergy complementing their knowledge, skills and experience. Apart from services and collaborations, postdoctoral and Ph.D. researchers carry out their research projects dedicated either to small-molecule or protein crystallography and structural biology.

Work in the Facility includes collaboration with other research groups and biotech/pharmaceutical companies, such as the WPD Pharmaceuticals, OncoArendi Therapeutics, Cellis or the Pharmaceutical Research Institute in Warsaw.

Moreover, we cooperate with **Dr. Sebastian Glatt** and **Dr. Przemysław Grudnik** (Structural Biology Core Facility, Jagiellonian University, Cracow) under the TT CF extension concerning on the commercial aspects (The Integrative Platform for Accelerated Drug Discovery – IPADD).

We are open to different forms of collaborations with individual researchers, research groups or biotech/pharma companies.

The project is supported by Foundation for Polish Science/European Union under the European Regional Development Fund (TEAM TECH CORE FACILITY/2017-3/4, POIR.04.04.00-00-31DF/17-00)

Figure 1

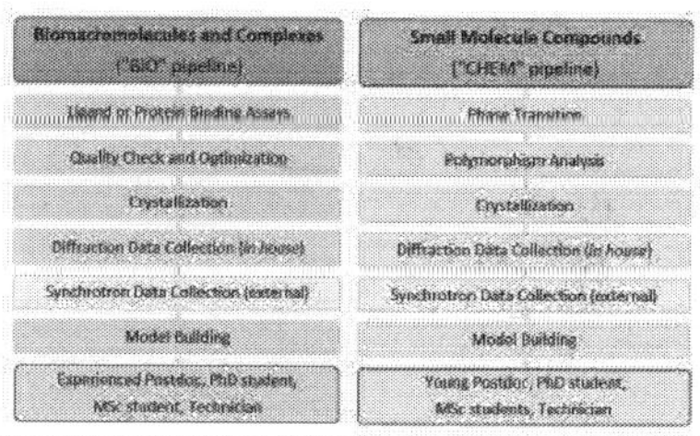

Fig. 1. Schematic view of the main pipelines of the Core Facility for Crystallography and Biophysics (CFCB).

P39

Structure-based Design of Broad-Spectrum Antivirals Targeting Coronaviruses and Enteroviruses: Peptidomimetic Cyanohydrins

L. Zhang[1,2], D. Lin[3], S. Anemüller[1], Y. Kusov[1,2], K. Rox[4,5], T. Kähne[6], G. Hansen[1], R. Hilgenfeld[1,2]
[1]University of Lübeck, Institute of Biochemistry, Lübeck, Germany
[2]German Center for Infection Research (DZIF), Hamburg - Lübeck - Borstel - Riems Site, Lübeck, Germany
[3]Changchun Discovery Sciences Ltd., Changchun, China
[4]Helmholtz Center for Infection Research, Braunschweig, Germany
[5]German Center for Infection Research (DZIF), Hannover - Braunschweig Site, Braunschweig, Germany
[6]Institute of Experimental Internal Medicine, Otto von Guericke University, Magdeburg, Germany

Despite the devastating consequences of emerging viruses such as Ebola virus or SARS coronavirus (SARS-CoV) for the affected patients, the development of antiviral drugs against these pathogens is not commercially viable in view of the comparatively low patient numbers. We are attempting to circumvent this problem for SARS-CoV and related coronaviruses (CoVs) by designing and developing antiviral drugs that target the large genus *Enterovirus* in addition to CoVs. Enteroviruses are responsible for many diseases such as Hand, Foot, and Mouth Disease, encephalitis, myocarditis, and poliomyelitis. The main protease of coronaviruses and the 3C protease of enteroviruses share a similar active-site architecture and a unique requirement for glutamine in the P1 position of the substrate, even though they are otherwise quite different. In order to obtain near-equipotent, broad-spectrum antivirals against coronaviruses and enteroviruses, we investigated the structure-based design of peptidomimetic cyanohydrins as inhibitors of main and 3C proteases. Six crystal structures of protease:inhibitor complexes were determined as part of this study, with the main proteases of SARS-CoV and MERS-CoV as well as the 3C protease of Coxsackievirus B3 (an enterovirus) as target enzymes. Compounds synthesized were tested against the recombinant proteases as well as in viral replicons and virus-infected cell cultures; none of them was cell-toxic.

Cyanohydrins were rarely reported as warheads for cysteine protease inhibitors. A crystal structure of a cyanohydrin inhibitor with enterovirus A71 3C protease has been described as a non-covalent complex (1), but we show here that different binding modes appear to be at work, depending on the identity of the viral protease and the stereochemistry around the cyanohydrin alpha-carbon.

Reference

[1] Zhai Y, Zhao X, Cui Z, Wang M, Wang Y, Li L, Sun Q, Yang X, Zeng D, Liu Y, Sun Y, Lou Z, Shang L, Yin Z.: *J. Med. Chem.* **58**, 9414-9420 (2015).

P40

Calcite manipulation by animals: the acicular-foliated microstructure of the limpets *Cellana toreuma* and *tramoserica* (Nacellidae, Patellogastropoda, Mollusca)

K. Berent[1], A. G. Checa[2]

[1]AGH University of Science and Technology, Academic Centre for Materials and Nanotechnology, Kraków, Poland

[2]University of Granada, Stratigraphy and Paleontology (Faculty of Sciences), Granada, Spain

Calcified biomaterials produced by invertebrates are used as a source of inspiration for the biomimics of novel materials. This is particularly the case of molluscs, which are the masters among metazoans in the production of both calcitic and aragonitic microstructures. Despite the few studies on limpets, it is clear that they secrete a variety of calcitic microstructures unparalleled by any other molluscan group. In *Cellana* there are two external calcitic layers. The outermost layer consists of elongated calcite granules and the second internal layer has a laminar microstructure hitherto undefined, made of imbricated thin folia (0.5-1 µm thick) which emerge on the internal shell surface in a clockwise direction as radially elongated laths. Folia are divided into very thin (<0.5 µm wide) acicle-like, arrow-pointed elements at a high angle to the growth margin. Internal to the calcitic layers, there is a thin aragonitic crossed-lamellar layer traversed by the myostracum.

In this way, with shell increase in diameter, folia elongate radially, whereas shell thickening is obtained by the clockwise extension of their wide growth margins. Introduction of new folia compensates for the increase in perimeter of the shell.

There is a sharp boundary between the outer granular and the more internal acicular-foliated layer, with the outer calcitic grains elongated in perpendicular to the growth surface drastically changing their morphology into the flat laths subparallel to this growth surface.

EBSD analyses demonstrate that the granular layer develops a stronger texture as the interface with the acicular-foliated layer is approached. The texture reaches a maximum in the latter layer. The changes in texture across the two layers are shown in Fig. 1 by means of EBSD texture plots.

This is another example of how animals are able to control their construction materials to fabricate particularly sophisticated and well-organized structures.

Figure 1

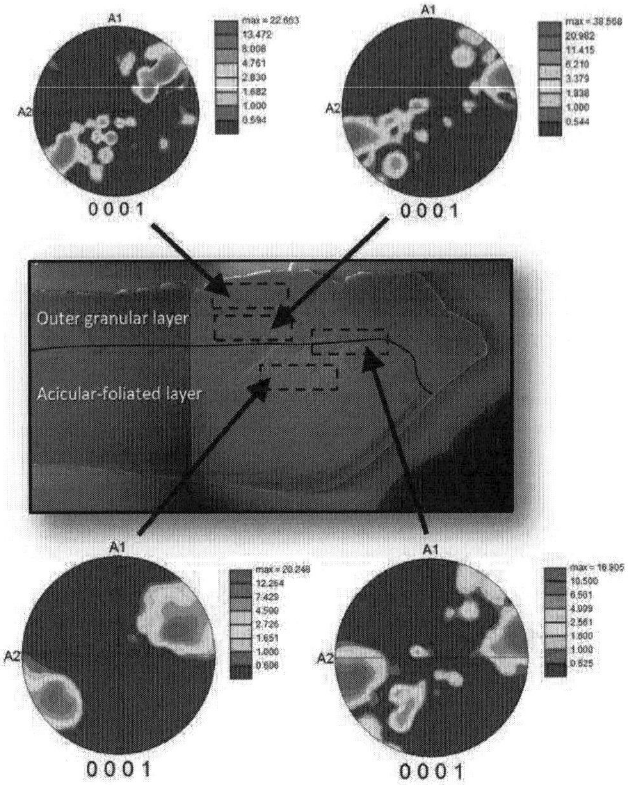

P41

Structural investigations of the fluorescence phenomenon in β-lactoglobulin complexes with hdaf

P. Wróbel[1], J. Loch[1], P. Bonarek[2], K. Lewiński[1]

[1]Jagiellonian University, Faculty of Chemistry, Kraków, Poland

[2]Jagiellonian University, Faculty of Biochemistry, Biophysics and Biotechnology, Kraków, Poland

Bovine β-lactoglobulin (BLG) is a protein belonging to the lipocalin family [1]. A characteristic feature of lipocalin structure is antiparallel β-barrel which is also a binding site for ligands [2]. The 5-(N-hexadecanoyl)aminofluorescein (HDAF) is fluorescent probe which can be used to detect hydrophobic cavities in the protein structures.

The aim of this project was to study fluorescence phenomenon of the HDAF in complexes with BLG mutants possessing substitutions in positions 39, 56, 58, 71, 92, 105 and 107. Selected BLG variants had different shape of the binding pocket and thus different affinity to aliphatic ligands.

BLG mutants were expressed in *E.coli*. Proteins were purified by anion exchange and size-exclusion chromatography. Crystals were obtained by the vapor diffusion method in the hanging drop setup. X-ray diffraction data were collected at XtaLAB Synergy (Rigaku). Structures of BLG-HDAF complexes were solved by molecular replacement.

The green fluorescence was observed for most of the determined structures indicating that the HDAF molecule was bound to the protein. Structures revealed that C16-aliphatic tail is bound in the hydrophobic part of the binding site, while fluorescein is located outside β-barrel and is often disordered. In some structures, the hydrolysis of HDAF was observed leading to dislocation of fluorescein to crystal water channels. In one structure the fluorescence quenching was observed probably due to strong HDAF interactions with protein.

The obtained crystal structures have shown that the enhancement or quenching fluorescent of HDAF depends on the shape of the binding side, so HDAF is a good candidate for site-specific fluorescent probe which could be used for determination of BLG mutants affinity to aliphatic compounds.

References

[1] Yongqin Wei, F. V. β-Lactoglobulin as a Nanotransporter for Glabridin. *ACS Omega* 3 (2018)
[2] Li Liang, M. Subirade, β-lactoglobulin. J. Phys. Chem 114 (2010)

P42

Structure of the c-di-AMP binding protein DarB

J. L. Heidemann[1], L. Krüger[2], J. Stülke[2], R. Ficner[1]
[1]Georg-August Universität Göttingen, Molecular Structural Biology , Göttingen, Germany
[2]Georg-August Universität Göttingen, General Microbiology, Göttingen, Germany

Bacteria are exposed to changes in nature for example a decreased or increased external osmolyte level. In order to adapt to these environmental changes and to transduce signals in the cell they utilize secondary metabolites like linear mononucleotides and cyclic dinucleotides. In recent years the second messenger cyclic di-AMP (c-di-AMP) attracted much attention. Not only because of its essentiality for the bacterial survival but also because of its complex involvement in different cellular processes and its diversity of interaction partners[1]. Up to the present time a plethora of c-di-AMP binding proteins are discovered. Many of these proteins are involved in potassium or osmolyte uptake[1]. c-di-AMP was identified to bind to conserved domains like the RCK_C domain or the CBS domain. Here we report three new crystal structures of the c-di-AMP binding protein DarB from *Bacillus subtilis* which consists of two CBS domains. All crystals diffracted to a resolution of 1.5 - 1.7 Å. Crystal were obtained by crystallizing DarB in presence of AMP, c-di-AMP or cyclic 3"3" cGMP-AMP. The crystal structures revealed two DarB monomers in the asymmetric unit forming a disk-like dimer. Surprisingly, the difference electron density map of each crystal suggested one of the described ligands in each of the supposed nucleotide binding site. In addition, we obtained a crystal form of apo DarB which led to a refined structure determined at 1.85 Å, which exhibits the same crystal packing ($P2_12_12_1$) and formation of a disk-like dimer in the asymmetric unit as the ligand bound structures of DarB.

[1] Gundlach J, Krüger L, Herzberg C, et al. Sustained sensing in potassium homeostasis: Cyclic di-AMP controls potassium uptake by KimA at the levels of expression and activity. *J Biol Chem.* 2019;294(24):9605-9614. doi:10.1074/jbc.RA119.008774

P43

Allosteric Modulation of cGAS/OAS Innate Immune Sensors: a Way to new Therapies

X. Zhou[1], O. Zeymer[1], P. Baruch[1], C. Goffinet[2], D. Manstein[1], R. Fedorov[3]
[1]Hannover Medical School, Institute for Biophysical Chemistry / Research Division for Structural Biochemistry, Hannover, Germany
[2]Center for Experimental and Clinical Infection Research, Hannover, Germany
[3]Hannover Medical School, Institute for Biophysical Chemistry / Research Division for Structural Biochemistry, Hannover, Germany

The cGAS/OAS family of innate immune sensors is among the most promising targets for the development of new antimicrobial and immunomodulatory agents. cGAS/OAS share highly conserved active sites with other nucleotidyltransferases. Targeting active sites of cGAS/OAS may interfere with other vital biological processes through unspecific inhibition (cross-reactivity). To avoid cross-reactivity and develop specific immunomodulatory drugs one can utilize allosteric mechanisms. Our work is focused on the identification, validation and targeting specific allosteric regulation sites in human cGAS/OAS. To achieve this end, we use the methodology that was developed by our group [1-3] and successfully applied to the related systems [3-4]. We were able to align chemical and structural cGAS states leading to the reconstruction of the conformational changes along the enzymatic cycle of human cGAS (hcGAS). This reconstruction revealed a previously unknown allosteric pathway connecting the catalytic center of hcGAS with the remote DNA-binding interface. The high-throughput *in silico* screening identified the molecular scaffolds for potential allosteric activity modulators of hcGAS. The experimental validation of their activity is now in progress. A similar approach is adapted to reconstruct the complete cycle of OAS. The first half of OAS cycle was reconstructed in our previous work [5]. The reconstruction of the complete OAS cycle will enable the discovery of specific allosteric modulation sites in OAS. The allosteric modulation of cGAS/OAS has potential for individualized medicine applications. Towards this end, we perform structure-function/activity relationship studies of naturally-occurring variations of human cGAS.

[1] Führing, et al. (2013) *ACS Catalysis.* 3:2976-2985.
[2] Führing, et al. (2015) *Sci. Rep.* 5:9618.
[3] Cramer, et al. (2018) *ACS Catalysis.* 8:2683-2692.
[4] Prakash, et al. (2019) *Molecules*, 24(5): 996.
[5] Lohöfener, et al. (2015) *Structure.* 23(5):851-862.

P44

Cyclic tetrapeptides as promising scaffold for innovative therapeutic agents: synthesis, crystallographic, biological and *in silico* studies

J. Bojarska[1], M. Remko[2], J. Wojciechowski[3], I. Madura[4], K. Kaczmarek[5], J. Zabrocki[5], M. Zimecki[6], W. Wolf[1]
[1]Łódź University of Technology, Department of Chemistry, Łódź, Poland
[2]Remedika 1, Bratislava, Slovakia
[3]Rigaku, Neu-Isenburg, Germany
[4]University of Warsaw, Department of Chemistry, Warsaw, Poland
[5]Łódź University of Technology, Department of Chemistry, Łódź, Poland
[6]Institute of Low Temperature and Structure Research, Polish Academy of Sciences, Wrocław, Poland

Cyclic tetrapeptides (CTPs) are becoming increasingly important in research of modern drugs development due to their unique structure and favorable features [1].

Our project focuses on the synthesis of new biologically active CTPs and holistic characterization of them. In this work, we will discussed selected representatives, coupled with CSD/PDB survey, keeping in mind knowing new biologically important supramolecular synthons and tectons in the relation to bio-complexes. In our laboratory, new CTPs (i.e. Pro-Pro-Phe-Phe containing β-2-isoproline, β-3–homoproline) were synthesized, which can result in breakthrough in therapy of patients with organ transplants. Additional studies revealed lack of toxicity and side effects as well as anti-inflammatory, immunosuppressive and skin anti-cancer activity [4]. Moreover, we investigated conformational, supramolecular and energy landscape of them. Our own research we combined with data from the CSD/PDB. We shed new light into potential of synthon idea [5] targeting into macromolecules due to better understanding of the bio-systems such as ligand–protein complexes [6]. Our preliminary studies revealed that in the CTP crystals mainly weak intermolecular interactions are involved in the formation of supramolecular H-bonding patterns, while interplay of them leads to cooperativity effect.

We hope that our findings will be of great value for advance in knowledge of supramolecular chemistry of biomolecules, which has a bright future ahead in new therapeutic approaches.

References

[1] Fosgerau 2015 *DrugDiscoveryToday*
[2] Frisch *et al* Gaussian09 2011
[3] Turner *et al* 2017 *CrystalExplorere17*
[4] Zabrocki *et al* 2011 **EP**2536732**US**2013/01086552013
[5] Desiraju 1995 *AngewChemInt;* Murkhejee 2015 *CrysGrowthDes*
[6] Bojarska, Kaczmarek, Zabrocki, Wolf 2018 *AdvancesInOrganicSynthesis* *Bojarska *et al* 2019 *JMolStruct* *Bojarska *et al* 2019 *NovApprDrugDesDev* *Bojarska *etal* 2019 *IntJNutrSci**Bojarska *et al* 2018 *WorldStructural&MolecularBiologyConference* Italy.

Structure-property-relationships

P45

Magneto-Structural Analysis in Metal-Organic Coordination Architectures Sustained by Hydrogen Bond: A CSD Study

K. Siddiqui[1]
[1]National Institute of Technology Raipur, Chemistry, Raipur, India

Introduction

In molecular magnetism hydrogen bond between metal complexes provides efficient mechanism for magnetic exchange couplings through supramolecular interactions such as hydrogen bridges has been reported for complexes based on diverse metal centers such as V(IV) , Cr(III) [, Fe(II) , Fe(III) , Co(II) ,, Ni(II) , and mixed-metal binuclear complexes. Copper chemistry is widely described experimentally and addressed theoretically in magneto-structural point of view especially with respect to hydrogen bonded binuclear systems.

Objectives

The article focuses on how the presence of hydrogen bond regulates the magneto-structural diversity in H-bonded metal-organic coordination complexes. The analysis was done by using the data retrieved by CSD search.

Materials and methods

CONQUEST- The Cambridge Crystallographic Data Centre was used to search and retrieve the information metallo-supramolecular networks from the Cambridge Structural Database (version 5.40 CSD update 2 - 2019).

Results

The Crystal Structure Database mining reveals the role of built-in H-bond, of coordinated ligands, towards the construction of {...O–Cu–O–H},{...O–H...O–Cu–N–N-H}, {...O–C–C–C–C–N–Cu–O–H}, {...O–C–O–Cu–O–H}, {...O–N–(O)2–Cu–O–H}, {...O–Cl–(O)2...H–N–Cu–N–H}, and {...O–C–O–Cu–N–N–H}, metal-organic supramolecular synthons which are instrumental in deciding the dimensionality and topology of metallo-supramolecular architecture and are pivotal in magneto-structural property relationship in copper coordination complexes.

Conclusion

Present study reveals the role of peripheral in-built hydrogen bonding groups incoordination complexes for construction of metal-organic supramolecular synthons and which in turnsprovides supramolecular platform for magnetic exchange pathways. These metal-organic supramolecular synthons play a significant role in context of crystal packing, structural diversity and magnetic exchange pathways.

Figure 1

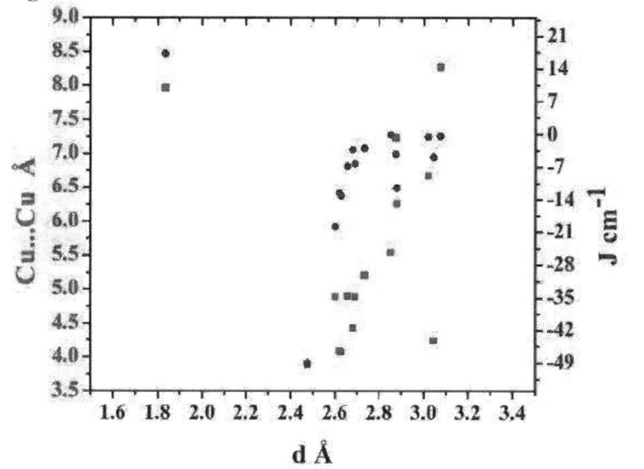

P46

Fluorescence and DFT Analysis in a Two-dimensional Zn^{II} Coordination Polymer

W. Wei[1], Y. An[1,2], M. Zhu[1], E. Ullrich[2]
[1]Shanxi University, Institute of Molecular Science, Taiyuan, China
[2]RWTH Aachen University, Institute of Inorganic Chemistry, Aachen, Germany

Based on the design principles of crystal engineering and structure-property relations, metal-organic frameworks (MOFs) with variable fluorometric sensing properties have been developed and well-documented in literature.[1-2] However, the judicial selection of organic ligands and metal centers plays an important role in the construction of functional MOFs. Especially, organic molecules containing conjugated π luminophores are widely applied as ligands in the fabrication of photoluminescent coordination assemblies.[3]

To prepare similar metal-organic frameworks (MOFs) with excellent fluorescence properties, the rigid organic ligand 2,2',3,3'-azobenzenetetracarboxylic acid (H_4abtc) containing a large conjugated π system was employed. In this ligand, the photochromic azo group can flip between its *cis* and *trans* isomers upon irradiation. By reacting the ligand with the d^{10} metal ion zinc^{II}, the novel 2D coordination polymer $[Zn(C_8H_3NO_4)(H_2O)_2]_n$ (**1**) was obtained under hydrothermal conditions. Interestingly, the solid-state fluorescence spectra of H_4abtc and (**1**) are similar both in intensity and emission wavelength. Following the Laporte law, the results of DFT and TDDFT calculation show the π–π* transition occurs from HOMO-2 (*au*) to LUMU (*ag*) and HOMO-1 (*au*) to LUMU (*ag*), which can explain the fluorescence mechanism.

Figure

Figure 1. (a) Reaction scheme to form the 2D MOF;(b) Solid-state emission spectra of H_4abtc (black) and (**1**) (red) at room temperature;(c) DFT energy levels for H_4abtc (left) and (**1**) (right).

Acknowledgements

The work is supported by China Scholarship Council, Shanxi University (scholarship for Yanyan An and Wenwen Wei) and the Natural Science Foundation of China (Grant No. 21571118 & 21671124).

References

[1] W. P. Lustig, S. Mukherjee, N. D. Rudd, *Chem. Soc. Rev.* 2017, *46*, 3242-3285.

[2] W. Liu, Y. L. Wang, Z. L.Bai, *ACS Appl. Mater. & Inter*faces. 2017, *9*, 16448-16457.

[3] M. Pan, W. M. Liao, S. Y. Yin, *Chem. Rev.* 2018, *118*, 8889-8935.

Figure 1

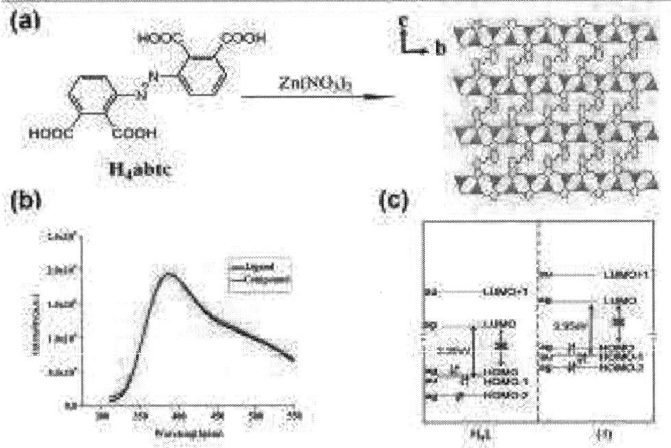

P47

Overcoming challenges in the synthesis of the Cu2ZnGe(S1-xSex)4 solid solution: development of a new synthesis route

S. Niedenzu[1,2], G. Gurieva[1], S. Schorr[1,2]

[1]Helmholtz-Zentrum Berlin für Materialien und Energie GmbH, Berlin, Germany

[2]Freie Universität Berlin, Berlin, Germany

The band gap energy variation in the compound semiconductors Cu2ZnGeSe4 (CZGSe) and Cu2ZnGeS4 (CZGS) from 1.4 eV to 1.7 eV, which is controlled by different S/(S+Se) ratios, renders the Cu2ZnGe(S1-xSex)4 solid solution (CZGSSe) an interesting material for multi-junction solar cells. In order to study structure-property relationship, a key-feature of materials design, bulk material is needed which is often synthesized by solid state reaction of pure elements. The system has a certain complexity due to the existence of different CZGS polymorphs, which crystallize in the tetragonal stannite as well as in the orthorhombic wurtz-stannite type structure. On the other hand CZGSe adopts the tetragonal kesterite type structure.

In this study polycrystalline Cu2ZnGe(S1-xSex)4 (x=0-1) mixed crystals were grown by solid state reaction using two newly developed approaches. In the first one binary compounds (CuS, CuSe, ZnS, ZnSe, and GeS) were used as starting materials. For deeper insights into the T-dependent phase transition in CZGS different temperature profiles have been chosen for the growth of S-rich CZGSSe mixed crystals. In the second approach, a mixture of binary compounds (CuS, CuSe, ZnS, ZnSe) and pure elements (Ge, S and Se) have been chosen to start with. The obtained material contain a single quaternary CZGSSe phase alongside secondary phases (WDX spectroscopy).

The bulk samples from the first approach are more inhomogeneous and show a higher amount of secondary phases. XRD analysis reveals that some S-rich CZGSSe mixed crystals show a coexistence of a tetragonal and an orthorhombic phase, both with the same chemical composition. The second approach leads to more homogenous samples with a lower amount of secondary phases. The quaternary CZGSSe mixed crystals adopt always the tetragonal crystal structure.

The advantages of one method over the other based on chemical composition, phase relations and a basic structural characterization of CZGSSe will be discussed.

P48

The determination of symmetry in $Sr_8MgR(PO_4)_7$ phosphates

D. Deyneko[1], S. Aksenov[2]

[1]Lomonoscov Moscow State University, Chemistry Department, Moscow, Russian Federation

[2]Lomonosov Moscow State University, Chemistry Department, Moscow, Russian Federation

A new series of solid solutions on $Sr_8MgEu_{1-x}Sm_x(PO_4)_7$ was synthesized and characterized. It is known that strontium phosphate crystallizes in the palmyrite-family mineral group with a centrosymmetric space group *R*-3*m*. However, according to published data, the introduction of a cation with a smaller ionic radius, such as Mg^{2+} or Zn^{2+}, into the structure of the cation leads to the formation of a related structure of the whitlockite mineral (β-$Ca_3(PO_4)_2$). These families of compounds are promising from the point of view of luminescent and ferroelectric properties.

The crystal structure of β-$Ca_3(PO_4)_2$ and related compounds contains 5 nonequivalent cationic sites *M*1 – *M*5, which differ in size and oxygen environment. It is also known that some kind of substitution, such as $Ca_8MgEu(PO_4)_7$, leads to the formation of a center of symmetry at position *M*3 and the equivalence of positions *M*1 and *M*2, thereby increasing the symmetry of the unit cell to space group *R*-3*c*. In general, it was found that such a transformation sp.gr. Positively affects the properties of substances. The isostructural phosphate $Ca_8MgEu(PO_4)_7$ crystallizes in *R*-3c with the parameters a = b = 10.36157 (6) Å, c = 37.09945 (4) Å, V = 3449.45 (4) $Å^3$, Z = 6. Calcium strontium substitution in the case of $Sr_3(PO_4)_2$, it leads to the formation of the *R*-3*m* palmyrite structure with parameters a = b = 5.388 (2) Å c = 19.7860 (1) Å, V = 497.44 (1) $Å^3$, Z = 3. Separation of these structures does not cause difficulties, since they are easily distinguishable by diffraction patterns.

In general, the choice of such a spatial group was confirmed by luminescent spectroscopy. The resulting substances exhibit intense red glow due to Eu^{3+} and Sm^{3+} cations contained in the crystal lattice.

The work was funded by Russian Science Foundation (Grant 19-77-10013).

Figure 1

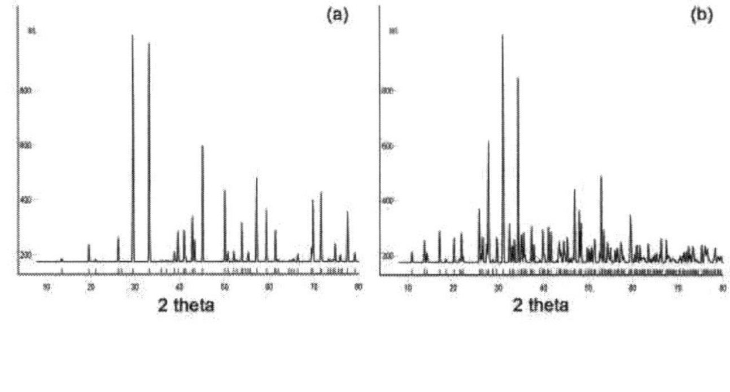

P49

Europium incorporation into $BaCa(CO_3)_2$

D. Spahr[1], L. Bayarjargal[1], V. L. Vinograd[2], B. Winkler[1]
[1]Goethe University Frankfurt, Institute of Geoscience, Frankfurt/Main, Germany
[2]Research Center Jülich GmbH, Institute of Energy and Climate Research (IEK-6), Jülich, Germany

Three naturally occurring polymorphs of $BaCa(CO_3)2$ (paralstonite, barytocalcite and alstonite) have been identified so far [1-3]. We synthesized a new polymorph and solved its structure [4]. The new polymorph is characterized by a small unit cell (V = 136 Å^3, Z = 1). The Ba and Ca cations are randomly distributed over the same Wyckoff position.

We studied barytocalcite and the $C2$ polymorph as a matrix material for the retention of actinides using Eu^{3+} as a proxy cation employing powder X-ray diffraction (PXRD), Raman and time resolved laser fluorescence spectroscopy (TRLFS) to demonstrate the Eu uptake into the solid. We found that adding Eu (< 1 %) during the barytocalcite synthesis leads to the formation of the $C2$ polymorph. The addition of Eu stabelizes the C2 phase relative to barytocalcite.

Furhtermore, Rietveld refinements on the $C2$ polymorph show a gradual decreases of the unit cell volume by ~1.5 % when the Eu concentration is increased to 6 % (Fig. 1). TRLFS shows a strong fluorescence with a life time of 570(50) µs for the $^5D_0$ to $^7F_2$ transition at ~616 cm^{-1} (Fig. 2), similar to incorporated Eu in calcite or aragonite [7].

TRLFS and PXRD imply the Eu-cation occupied the Ba/Ca position, although it has a much smaller cation radius. Our results demonstrate that the $C2$ polymorph can accommodate up to ~ 6 mol% of Eu and therefore may be a promising matrix material for actinides in contrast to calcite.

The authors are grateful for funding by DFG (Wi1232-401).

[1] B. Dickens and J. S. Bowen, *J. Res. Nat. Bur. Stand.*, 75A:197–203, 1971.
[2] H. Effenberger, *N. Jb. Miner. Mh.*, 353–363, 1980.
[3] F. Sartori, *Lithos*, 8:199–207, 1975.
[4] D. Spahr *et al.*, *Acta Cryst.*, B75:291–300, 2019.
[5] C. Fruhner et al., *Eur. J. Mineral.*, 30:711–720, 201.

Fig. 1: Relative unit cell volume V/V_0 (b) as a function of the nominal Eu concentration in $(BaCa)_{(2-x)}Eu_x(CO_3)_2$.

Fig. 2: Life time of the $^5D_0$ to $^7F_2$ transition at ~616 cm^{-1} in $(BaCa)_{1.94}Eu_{0.06}(CO_3)_2$.

Figure 1

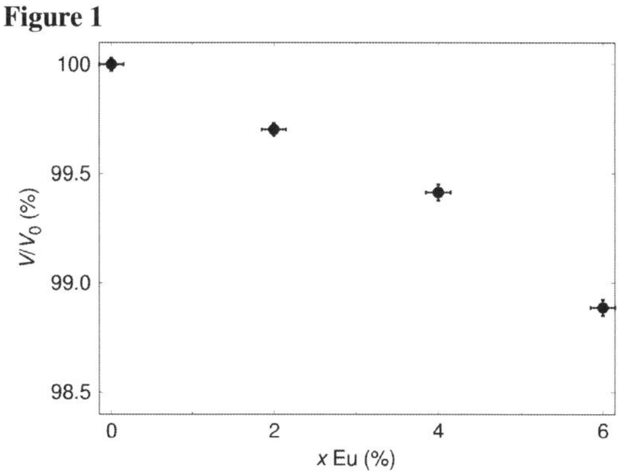

Figure 2

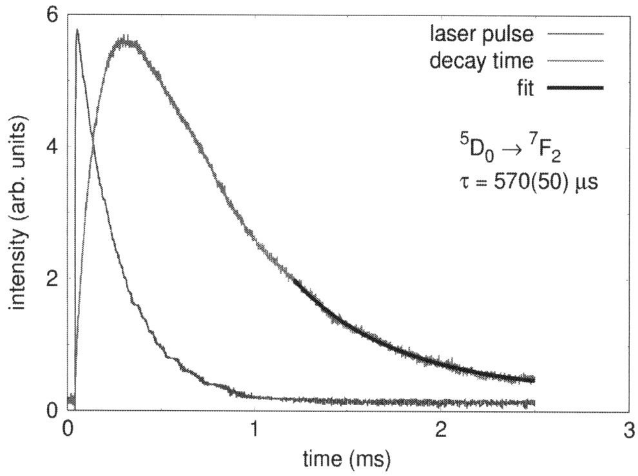

P50

Investigating the morphology of nanostructured mixed metal oxides (Ir/TiO_x) and its impact on the electrocatalytic OER-activity

J. Marquardt[1], S. Kiske[2], M. Bernicke[2], H. M. H. Raza[3], N. Pinna[3], R. Kraehnert[2], F. Emmerling[1,4]
[1]Federal Institute for Materials Research and Testing (BAM), Structural Analysis, Berlin, Germany
[2]Technische Universität Berlin, Institut für Chemie, Berlin, Germany
[3]Humboldt-Universität zu Berlin, Institute of Chemistry, Berlin, Germany
[4]Humboldt-Universität zu Berlin, Berlin, Germany

The electrocatalytic conversion of water into molecular hydrogen and oxygen under the utilization of excess renewable energies, such as wind power [1], photovoltaics [2] and hydroelectric power [3] is one possible pathway to establish a sustainable hydrogen economy. The obtained hydrogen is either stored and used in a fuel cell or consumed on-site in industrial applications Water electrolysis systems (WES) are based on two half cell reactions, such as oxygen evolution reaction (OER) and hydrogen evolution reaction (HER) which both proceed simultaneously [4]. The OER suffers from slow reaction kinetics and thus limits the overall performance [5]. The most promising compounds in acidic electrolysis are IrO_2 and RuO_2 [5]. Due to their rare abundance and extremely high price a wide use of acidic WES was prevented. Lowering the catalysts noble metal content by mixing iridium with titanium [6] reduces the production costs. Thin films are produced by dip coating a solution of metal oxide precursors alongside with a polymer template dissolved in ethanol. The obtained samples are subsequently calcined to the remove the template and adjust crystallinity. Finally, an additional iridium deposition step was performed on the outer surface plane area. Understanding the influence of structural and morphological aspects on the OER-activity is beneficial to further optimize WES. The current presentation will thus give detailed insights to structural aspects obtained by Raman spectroscopy, small- and wide-angle X-ray scattering which are then combined with electrochemical parameters to deduce structure-activity relationships. [1] A. Costa et al., *Renewable Sustainable Energy Rev.* **2008**, 12,1725–1744;[2] A. J. Frank et al., *Coord. Chem. Rev.* **2004**, 248,1165–1179;[3] R. Bakis, *Energy Sources Part B* **2007**, 2,259–266;[4] D. Bernsmeier et al., *ChemElectroChem* **2017**, 4, 221-229;[5] M.Carmo et al., *Int. J. HydrogenEnergy* **2013**, 38, 4901;[6] D. Bernsmeier et al., *ChemSusChem* **2018**, 11, 2367-2374

P51

The kesterite – stannite structural phase transition: comparison of the $Cu_2(Zn,Fe)S_4$, $Cu_2(Zn,Cd)S_4$ and $Cu_2(Zn,Mn)SnSe_4$ series

S. Schorr[1,2], S. Niedenzu[1,2], A. Franz[2], G. Gurieva[2]
[1]Freie Universität Berlin, Institute of Geological Sciences, Berlin, Germany
[2]Helmholtz-Zentrum Berlin für Materialien und Energie GmbH, Structure and Dynamics of Energy Materials, Berlin, Germany

Alloying of cations or anions is an established method for tailoring the band gap energy in compound semiconductors. In the presented work we discuss the solid solution series $Cu_2(Zn,Fe)S_4$, $Cu_2(Zn,Cd)S4$ and $Cu_2(Zn,Mn)SnSe_4$ in which divalent Zn is substituted systematically by Fe, Cd and Mn. The end members Cu_2ZnSnS_4 and $Cu_2ZnSnSe_4$ crystallize in the kesterite structure (s. g. I-4) [1], whereas the end members Cu_2FeSnS_4 [1], Cu_2CdSnS_4 [2] and $Cu_2MnSnSe_4$ [2] crystallize in the stannite structure (s. g. I-42m). Both the kesterite and the stannite structure are based on a tetrahedral coordination but different cation arrangements: perpendicular to the crystallographic *c*-axis in the stannite structure Zn-Sn planes change with Cu-Cu planes, whereas in the kesterite structure Cu-Sn planes and Cu-Zn planes are present.

We studied the structural transition from the kesterite to the stannite structure in these three solid solution series as well as the variation of the band gap energy with the chemical composition. The detailed structural investigations are based on neutron powder diffraction experiments which provide the cation distribution in the crystal structure allowing a distinction between the isoelectronic cations Cu^+ and Zn^{2+} as well as the electronic similar cations Mn^{2+} and Fe^{2+}.

A remarkable structural feature is the abrupt change of the tetragonal deformation *c/2a* (*a* and *c* are the lattice parameter). In the series $Cu_2Zn_{1-x}Fe_xS_4$ and $Cu_2Zn_{1-x}Cd_xS_4$ this change occurs at the same x value of x~0.3. In Cu_2ZnSnS_4 is c/2a~1, where as in $Cu_2ZnSnSe_4$ is c/2a<1. Thus in the $Cu_2(Zn,Mn)SnSe_4$ series c/2a increases by substituting Zn by Mn until a value of c/2a~1 is reached. The structural kesterite-stannite transition is characterized by a cation re-distribution process which will be discussed in detail.

[1] Schorr et al., Europ. J. Mineral. 19 (2007) 65
[2] Schäfer et al., Z. Krist. 145 (1977) 356

P52

The kinetics of the Norrish-Yang reaction in crystals – X-ray diffraction and Raman spectroscopy studies

K. Konieczny[1], A. Szczurek[2], J. Bąkowicz[1], I. Turowska-Tyrk[1]
[1]Wrocław University of Science and Technology, Advanced Materials Engineering and Modelling Group, Wrocław, Poland
[2]Wrocław University of Science and Technology, Department of Mechanics, Materials Science and Engineering, Wrocław, Poland

The Norrish-Yang reaction is an intramolecular process leading to a four-membered ring (Figure 1). Its path depends strongly on a crystal structure. A very interesting feature is that the proceeding reaction modifies the crystal structure and in this way influences its further path, what is the subject of our studies. This means that we analyze the reaction kinetics basing not only on the crystal structure before UV irradiation, *i.e.* for the initial conditions of the reaction, but also basing on many partly reacted structures.

Figure 1. The Norrish-Yang reaction for the studied compound.

We studied single crystals of hexamethylenetetramine 4-(2,4,6-triisopropylbenzoyl)benzoate at ambient and high pressures. The photochemical reaction was induced by means of UV-vis radiation in a stepwise manner. After each step the X-ray crystal structure was determined and the Raman spectrum was recorded.

We successfully conducted the Norrish-Yang reaction of the studied compound until 100% conversion of the reactant. The Johnson-Mehl-Avrami-Kolgomorov (JMAK) model was applied in order to describe the reaction kinetics [1, 2]. It was revealed that the reaction proceeded in two steps. In the first stage the reaction was homogenous with constant rate and in the second stage its autoinhibition was observed. The reasons of this phenomenon were the structural changes along with the reaction progress in: (i) the geometry of the reaction center, (ii) C-H···π interactions, (iii) N-H···O hydrogen bond modifying the character of the carboxylate group. This was consistently indicated by crystal structures and Raman spectra.

[1] J. Bąkowicz, J. Olejarz, I. Turowska-Tyrk, J. Photochem. Photobiol. A: Chem. (2013) 273, 34-42.
[2] T. Runčevski, M. Blanco-Lomas, M. Marazzi, M. Cejuela, D. Sampedro, R. E. Dinnebier, Angew. Chem. (2014) 126, 6856-6860.

Figure 1

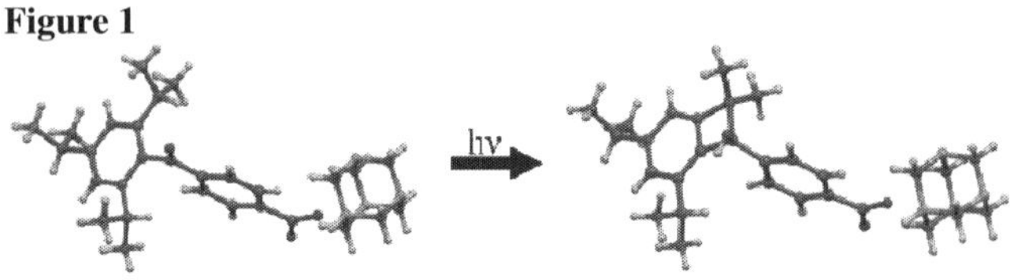

P53

Application of violuric acid as a possible optical sensor for basic amino acids detection: Crystal structures and spectroscopic studies

A. Rydz[1], M. Gryl[1], K. M. Stadnicka[1]
[1]Jagiellonian University, Faculty of Chemistry, Kraków, Poland

Introduction

Selective detection of proteins, peptides and amino acids is very important as these substances are fundamental in biological processes. Sensing of amino acids is carried out using combined spectroscopic, electrochemical or chromatographic techniques what leads to time-consuming and sometimes very complex procedures.[1-4]

Objectives

Our studies were focused on finding simple and efficient method for basic amino acids detection utilizing the capability of violuric acid (VA) to form colored salts with organic co-formers. VA was considered as a hypothetical sensor for detection of basic amino acids: histidine (HIS), lysine (LIZ) and arginine (ARG).

Materials and methods

Crystal structures of VAHIS and VALIZ salts were determined by single-crystal X-ray diffraction studies and characterized by spectroscopic measurements in the solid state. UV-Visible titration was carried out for the aqueous solutions of VA in the absence and presence of the various concentrations of HIS, LIZ or ARG.

Results
UV-Vis titration was used to define the detection limits for the absorption maximum at $\lambda = 544$ nm in the case of all studied solutions. Due to different interactions between components in the crystalline phases, the particular basic amino acid could be distinguished based on the different position of absorption maxima in the solid state.

Conclusions
The basic amino acids can be detected in solution and selectively distinguished in the solid state by UV-Vis spectroscopy. The utilization of violuric acid as sensor of basic amino acid proved to be simple method based on standard reaction of salt formation without any necessity to apply additional reactants or procedures.

[1] Altaf, A. A. *et al. R. Soc. Open sci.* 3, 2016, 160351
[2] Luque, G. L. *et al. Talanta* 71, 2007, 1282-1287
[3] Le Boucher, J. *et al. Clinical Chemistry* 43:8, 1997, 1421-1428
[4] Ijeri, V. *et al. Electroanalysis* 19, 4, 2007, 510-514

P54

Local structure, dynamics, and expansivity of α polyamide 6 through the glass transition

M. Terban[1], S. Bette[2], B. Hinrichsen[3], P. Desbois[3], M. Etter[4], A. Schökel[4], R. E. Dinnebier[2]
[1]Brookhaven National Laboratory, Condensed Matter Physics and Materials Science, Upton, United States
[2]Max-Planck-Institute for Solid State Research, Scientific Facility X-ray diffraction, Stuttgart, Germany
[3]BASF, Ludwigshafen, Germany
[4]Deutsches Elektronen Synchrotron (DESY), Hamburg, Germany

Detailed structural assessment of nonoriented semicrystalline polymers is often limited due to the high degree of amorphous content and structural disorder. This makes the determination of the structural response to different stimuli difficult, because diffraction data lead to ambiguous patterns which are difficult to index quantitatively, or ascribe to specific structural phenomena. In recent work, we demonstrated that the pair distribution function (PDF) obtained from total scattering measurements provides suitable information for obtaining quantitatively reliable lattice parameters in nanocrystalline polyamides [2]. In this study, we use pair distribution function analysis of total scattering data collected at Petra III at DESY to quantify the local structure and dynamics of polyamide 6 as it is heated through the glass transition. Structural refinements allow us to track the directionally dependent expansion behavior of the nanocrystallites in the different temperature ranges.

The recent implementation of an in situ Raman laser with our in-lab diffractometer allowed us to gain further insight into the changes in intra- and intermolecular dynamics. Coupled with the PDF analysis, these methods allow us to gain further insight into the molecular origins of thermal expansion behavior in nonoriented, semicrystalline polymers.

[1] Terban, M. W.; Pütz, A.; Heinemeyer, U.; Hinrichsen, B.; Desbois, P.; Dinnebier, R. E.; submitted, 2019

P55

Ex-situ and *in-situ* X-ray diffractometry as the applicable tool to determine structural properties and cycling performance of P2-$Na_{0.67}Mg_yMn_{1-y}O_2$ (y=0.1, 0.2, 0.3) cathodes for Na-ion batteries

G. Ważny[1], K. Walczak[1], A. Kulka[1], J. Molenda[1]
[1]AGH University of Science and Technology, Faculty of Energy and Fuels, Department of Hydrogen Energy, Kraków, Poland

Layered transition metal oxides gained their popularity as cathode materials for Na-ion batteries due to their structural stability during intercalation and deintercalation. Sodium ions are placed in van der Waals gaps between transition metal layers, therefore their movement does not affect the crystal structure significantly. Among others, Mn-based layered oxides are widely investigated [1].

The main objective of this work was the study of $Na_{0.67}Mg_yMn_{1-y}O_2$ (y=0.1, 0.2, 0.3), especially concerning its crystal structure, phase transitions during the cell performance and in equilibrium, as well as determiation of the most favourable synthesis conditions.

$Na_{0.67}Mg_yMn_{1-y}O_2$ (y=0.1, 0.2, 0.3) were prepared via solid state reaction. Crystal structure of as-synthesized materials, as well as *ex-situ* and *in-situ* measurements, was determined based on X-ray diffraction. Synthesized materials were used to prepare cathodes for sodium cells, which were studied for refering their cycling performance to their crystal structure, sodium content and magnesium substitution.

Optimization of sintering procedure allowed to obtain single-phase hexagonal structure, characterized as *P6₃/mmc* space group. Proper selection of Mg substitution, electrolyte type and cycling voltage range resulted in receiving the cell that reaches over 170 mAh/g with very good capacity retention. XRD studies indicate that the material preserves its structure with at least 0.5 mol Na.

XRD is a significant tool for the studies on cathode materials. Especially *ex-situ* and *in-situ* measurements allow to obtain a complex view on structural properties in the equilibrium state, as well as during the cell performance.

Acknowledgements
This work was funded by National Science Centre, Poland under grant no. 2018/31/N/ST8/01662.

References
[1] Talaie E., Duffort V., Smith H.L., Fultz B., Nazar L.F., Energy Environ. Sci., 8 (2015) 2512-2523

P56

Stability of baclofen crystal forms related to ɣ-lactam impurity formation

M. Główka[1], W. Maniukiewicz[1]
[1]Łódź University of Technology, Institute of General and Ecological Chemistry, Łódź, Poland

Baclofen shows high spasmolitic activity and is used mostly in the treatment of multiple sclerosis, spiral cord injuries and disease, or cerebral palsy [1-3]. Unfortunately, this pharmaceutically active substance is prone to cyclization reaction with formation of a 5-membered ɣ-lactam (Scheme 1). The rearrangement was first observed for racemic Baclofen at elevated temperature (above

180°C) [4]. The main objective of this study was to explain the very high susceptibility of crystalline form B of enantiomeric (R)-Baclofen to cyclization reaction that already occur at approx. 60°C. For this purpose the crystal structure of form B of (R)-Baclofen was determined and compared with the structure of (DL)-Baklofen [5]. The study showed that the analyzed structures at the Cβ atom adopt two essentially different crystal conformations - bent or expanded. The conformation in this case can be best characterized by N...C_x (C_x atom in the carboxyl group) distances, below 3.5Å for the enantiomer or more than 4Å for the racemic crystals, respectively. This feature may be responsible for the lower stability of (R)-Baclofen in the cyclization reaction.

Scheme 1. The formation of the ɣ-lactam by cyclodehydration (lactamization) reaction

References

[1] Pedersen, E., Arlien-Soborg, P., Grynderup, V., and Henrikson, O. (1970), Acta Neurol. Scand, 46, 257.
[2] Hudgson, P., and Weightman, D. (1971), British Medical Journal, 4, 15.
[3] Korsgaard, S. (1976), Acta Psychiat. Scand. 54, 17.
[4] Borka L., (1979), Acta Pharm Suec, 16, 345.
[5] Maniukiewicz, W., Oracz,M., Sieron, L., (2016), Journal of Molecular Structure, 1123, 271

Figure 1

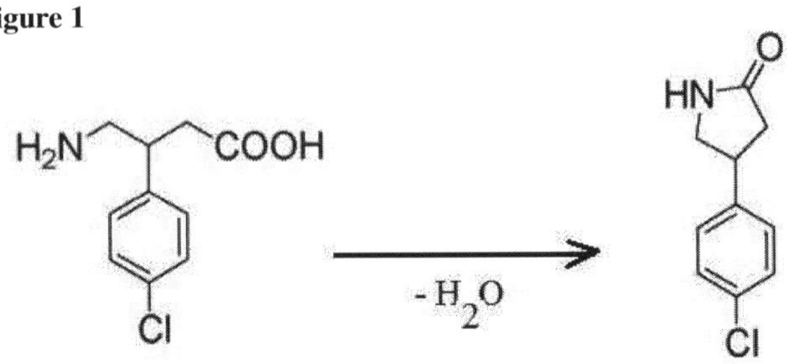

P57

Operando XRD studies during electrochemical intercalation of $Na_{0.66}Li_{0.22}Ti_{0.78}O_2$ – P2-type anode material for Na-ion batteries.

M. Nowak[1], W. Zając[1], A. Kulka[1], J. Molenda[1]
[1]AGH University of Science and Technology, Faculty of Energy and Fuels, Department of Hydrogen Energy, Kraków, Poland

One of the challenges that preclude a broader application of large scale energy storage systems is to design materials that need to withstand thousands of discharging/charging cycles. $Na_{0.66}Li_{0.22}Ti_{0.78}O_2$ (NLTO) is an attractive P2-type anode material for such applications due to its low working potential, high stability and minor volume changes during cycling. However, a mechanism of sodium intercalation and diffusion needs to be better understood.

NLTO powders were prepared via citrate sol-gel method starting from $NaNO_3$, $LiNO_3$ and titanium isopropoxide with final heat treatment at 850°C for 2h. X-ray diffraction was employed to examine phase purity and Rietveld refinement was used to calculate lattice parameters. Electrodes for in-situ XRD experiments were prepared by mixing NLTO powder, KetjenBlack, and PVDF in a 70:20:10 ratio. Totally, 160 XRD scans in 10-60°C range, each 30 min long were gathered during electrochemical cycling to evaluate structure changes.

All diffraction peaks can be indexed with *P6₃/mmc* space group typical for P2-type layered oxides. The structure consists of $[Ti/Li]O_6$ octahedra sharing edges and corners which consecutively form $[Ti/Li]O_2$ slabs with "ABBA" oxygen stacking. Between these slabs, sodium ions occupy two prismatic Wyckoff positions: 2b – sharing edges and 2d – sharing faces with $[Ti/Li]O_6$ octahedra. Lattice parameters were calculated to be a = 2.9639 and c = 11.1309 Å. Sodium intercalation is accompanied by a monotonic shift of (00l) peaks to lower angles and (h00) peaks to higher angles. No new reflections were observed what suggest stability of the P2 phase in 0.01-2.5 voltage range. Very low volume change of 1.4% can explain low capacity fade during further cycles, whereas a change of occupancy of 2d and 2b positions may account for diminishing sodium diffusion coefficient with a depth of discharge. Finally, operando XRD enabled to point areas where KetjenBlack"s contribution to calculated capacity is considerable.

Figure 1

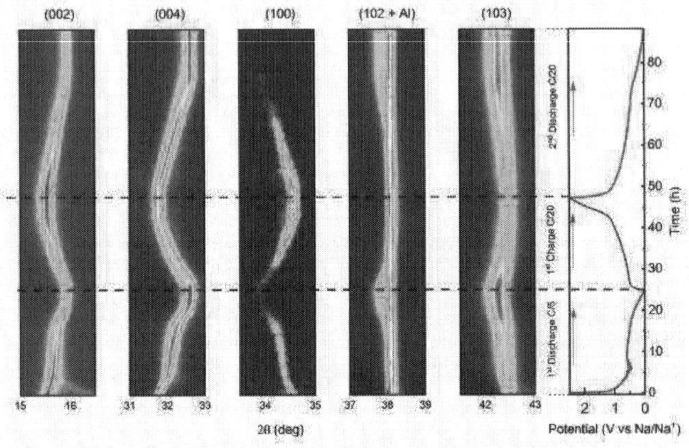

P58

How *in house* X-Ray Total Scattering can help gain a deeper insight into Solid-state Electrolytes

A. L. Hansen[1], C. Fritsch[1], M. Knapp[1], H. Ehrenberg[1]
[1]Karlsruhe Institute of Technology, Institute for Applied Materials - Energy Storage Systems (IAM-ESS), Eggenstein-Leopoldshafen, Germany

Introduction

An alternative to classical battery devices are all-solid-state batteries (ASSB), promising better safety and superior performance. To develop highly performing ASSBs, solid-state electrolytes with superior ion mobility are crucial. Especially glasses, with their enhanced bulk conductivity and lowered interfacial resistance, show encouraging properties. Of course, due to the amorphous nature of glasses, it is difficult to establish a connection between their structure and these properties.

Objectives

Based on total scattering measurements, pair distribution function (PDF) analysis fills the gap between very local techniques (NMR, IR, Raman) and methods giving us an understanding of the global crystallographic structure (XRD). While spectroscopy helps in detecting different polyhedral species, we need PDF to clarify their medium range order, *i.e.* the length and strength of their coherence.

Methods

We perform *in house* total scattering experiments using a STOE Stadi P powder diffractometer with Ag-Kα1 radiation equipped with two Dectris MYTHEN 1k detectors.

Results

In first investigations on Li based thiophosphates, we were able to identify changing building blocks during amorphization and recrystallization.1 Presence and coordination of these building blocks have influence on the ionic conductivity and therefore need to be included into models and calculations.

Conclusion

Without dispute, the best method to perform total scattering measurements on glasses, is the use of a high energy synchrotron source. But beamtime at international facilities is highly competitive and restricted to a few days per cycle. Our results indicate that *in house* total scattering measurements too can give valuable insights into the structure of glasses.

[1] Stöffler, H. *et al.* Amorphous versus Crystalline Li 3PS4 : Local Structural Changes during Synthesis and Li Ion Mobility. *J. Phys. Chem. C* 123, 10280–10290 (2019).

P59

Structure and biological activity of selected amidrazone derivatives

A. Olczak[1], K. Gobis[2], I. Korona-Głowniak[3], M. Krause[2], I. Mazerant-Politowicz[1], M. Szczesio[1]
[1]Łódź University of Technology, Institute of General and Ecological Chemistry, Łódź, Poland
[2]Medical University of Gdansk, Department of Organic Chemistry, Gdańsk, Poland
[3]Medical University of Lublin, Department of Pharmaceutical Microbiology, Lublin, Poland

Tuberculosis is caused by intracellular pathogen *Mycobacterium tuberculosis* [1]. As a result of widespread use of antibiotics some strains of this pathogen derived immunity and for that reason new tuberculostatic agents are sought after. Amidrazone derivatives (Scheme) are one of the promising groups of compounds in this regard. Their activity against tuberculosis and other bacteria was tested. Single crystal diffraction analysis was used to determine the molecular and crystal structure of the studied compounds. Interesting feature of their structure is possibility of tautomerism, which results in a variety of hydrogen bonds patterns and various types of packing of molecules in the crystal state. In addition, in the zwitterionic form (in comparison with the nonpolar form) different intramolecular hydrogen bonds can be formed: NH...N imposing a specific conformation of the piridine ring and NH...S stabilizing the flat conformation of the molecule.

[1] L.M. Fu, C.S. Fu-Liu, Tuberculosis 82, 85, (2002)

Funding for this research was provided by the National Science Centre (Poland), decision number DEC-2017/25/B/NZ7/00124.

Figure 1

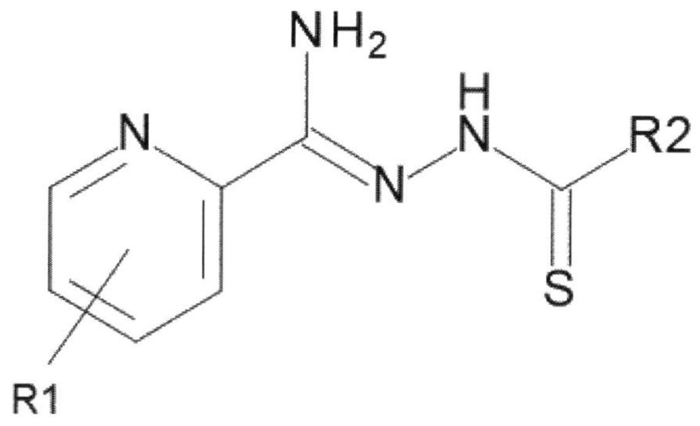

P60

Structure and antibacterial activity of selected hydrazide derivatives

M. Szczesio[1], K. Gobis[2], I. Korona-Głowniak[3], D. Zięmbicka[2], A. Olczak[1]
[1]Łódź University of Technology, Institute of General and Ecological Chemistry, Łódź, Poland
[2]Medical University of Gdansk, Department of Organic Chemistry, Gdańsk, Poland
[3]Medical University of Lublin, Department of Pharmaceutical Microbiology, Lublin, Poland

In 2016, over 10 million people fell ill with Tuberculosis [1]. 1.7 million died from the disease. Tuberculosis (TB) is one of the top 10 causes of death worldwide. Hydrazides derivatives (scheme) are widely studied in terms of their biological and particularly antibacterial activity. In this work their activity against tuberculosis and other bacteria was tested. Single crystal diffraction analysis was used to determine the molecular and crystal structure of the selected hydrazide derivatives.

Overall conformations of molecules of the studied compounds depend mainly on the type of R3 substituent (scheme). Monothioester derivatives are characterized by large conformational twist around N N bond with C N N C torsion angle between 87° and 180°.

All studied compounds show rather low activity against *Mycobacterium tuberculosis*, but their activity against the other Gram-positive bacteria tested in this study is quite significant.

[1] http://www.who.int/tb/en/

This project was funded by the National Science Centre (Cracow, Poland) on the basis of decision number DEC-2017/25/B/NZ7/00124.

Figure 1

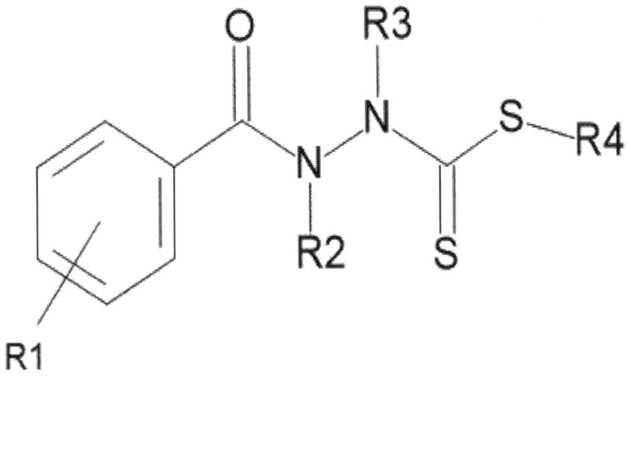

P61

Influence of nitro substituent in phenylpiperazine derivatives of 3-methyl-5-spiro(fluorene)hydantoin on the crystal and molecular structures

E. Zeslawska[1], W. Nitek[2], J. Handzlik[3]
[1]Pedagogical University, Kraków, Poland
[2]Jagiellonian University, Kraków, Poland
[3]Jagiellonian University , Medical College, Kraków, Poland

According to WHO, cancer is still a major human health problem leading to death 9.6 million people in 2018 [1]. There is a great interest in experimental chemotherapy regarding multidrug resistance (MDR) inhibitors and new anticancer agents. The cellular overproduction of MDR efflux pumps, for example ABCB1 (P-glycoprotein, P-gp), is one of the more relevant mechanisms underlying MDR, which is capable to promote the efflux of cytotoxic drugs out of the cells [2]. Our previous studies [3,4] allowed to identify a series of 5-spirofluorenehydantoin derivatives, which caused a potent inhibition of P-gp. The more active 3-methyl-5-spirofluorenehydantoin derivatives were chosen for further modification in order to improve their biological activity.

In term to extend knowledge about pharmacophoric features responsible for ABCB1 inhibitory properties of arylpiperazine derivatives of 3-methyl-5-spirofluorenehydantoin, we have prepared and performed the crystal structure analyses for 1'-[3-(4-phenylpiperazin-1-yl)propyl]-3'-methyl-spiro(fluorene-9,5'-imidazolidine)-2',4'-dione (**1**), 1'-[5-(4-phenylpiperazin-1-yl)pentyl]-3'-methyl-spiro(fluorene-9,5'-imidazolidine)-2',4'-dione (**2**) and their nitro derivatives (**1n** and **2n**).

Both three and five carbons linkers adopt a bent conformation that does not change after introduction of nitro substituent (Fig. 1). There is a significant difference in the arrangement of piperazine ring with respect to the linker and in the mutual orientation of piperazine and aromatic rings. The differences are also visible in the arrangement of 5-spirofluorenehydantoin with respect to the linker.

Figure 1. Superposition of (a) **1** and **1n** and (b) **2** and **2n** with respect to the linker. Colour key: **1** and **2** grey; **1n** and **2n** light grey.

References

[1] WHO Media centre, Fact sheets, Cancer, September 2018.
[2] C. Hegedüs, K. Truta-Feles, G. Antalffy *et. al. Biochem Pharmacol.* 84 (2012) 260.
[3] E. Żesławska, A. Kincses, G. Spengler *et al. Bioorg. Med. Chem.* 24 (2016) 2815.
[4] E. Żesławska, A. Kincses, G. Spengler *et al.Chem. Biol. & Drug Des.* 93 (2019) 844.

Figure 1

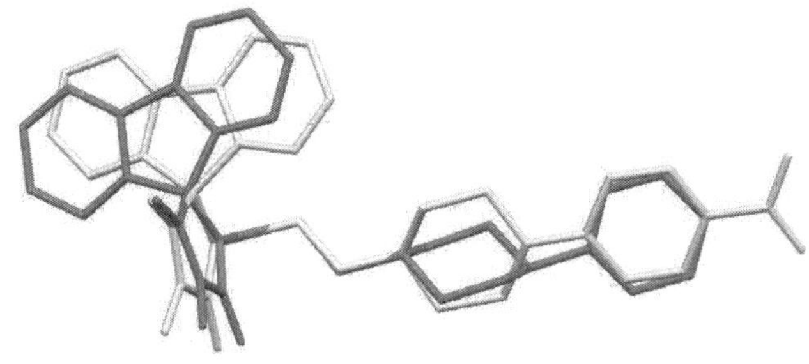

Figure 2

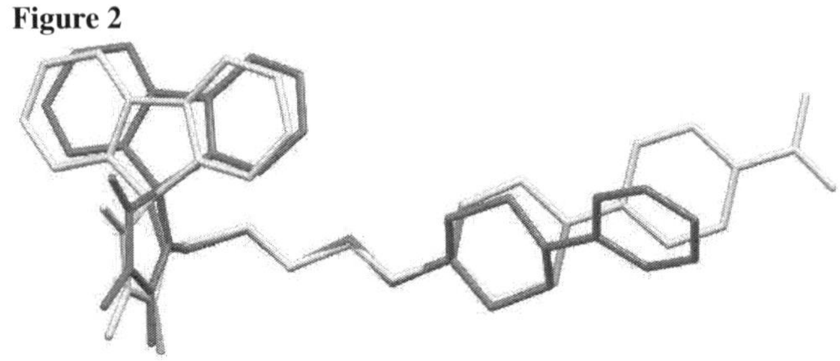

P62

Effect of Sc^{3+} and Al^{3+} Ion Doping on Magnetic Properties of Nd-stabilized SrM-Hexaferrite Nanostructures

A. Hilczer[1], K. Pasińska[2], A. Pietraszko[2], B. Andrzejewski[1]
[1]Polish Academy of Sciences, Institute of Molecular Physics, Poznań, Poland
[2]Institute of Low Temperature and Structure Research, Polish Academy of Sciences, Wrocław, Poland

M-hexaferrites are of great interest due to their magnetoelectric and magnetic quantum paraelectric behavior related to an exceptional role of FeO_5 bipyramid [1,2]. 24 ferric ions arranged in $P6_3/mmc$ space group with Z=2 in form of ferrimagnetic along the *c*-axis are due to superexchange Fe^{3+}-O^{2-}-Fe^{3+} interactions. Their properties are influenced also by the microstructure. The interactions can be modified by substitution of the ferric ions with different ionic radii and/or valence. We were interested in the effect of Sc^{3+} and Al^{3+} doping on the magnetic response of $Sr_{0.95}Nd_{0.05}Fe_{12-x}B_xO_{19}$.

$Sr_{0.95}Nd_{0.05}Fe_{12-x}Sc_xO_{19}$ and $Sr_{0.95}Nd_{0.05}Fe_{12-x}Al_xO_{19}$ nanocrystallites were obtained by citric sol-gel method (agglomerated platelets 50-100 nm thick) and hydrothermal synthesis (10-20 nm thick platelets), respectively. Single M-hexaferrite phase was confirmed in all samples by XRD studies (X"Pert PANalytical, CuK_α, Rietveld refinement). Lattice parameters were found to increase with Sc concentration, whereas Al-doping did not influence the parameters. Magnetic measurements (PPMS Quantum Design, 10-300 K) showed a decrease in specific magnetization after doping and a softening of magnetic properties. Longitudinal conical magnetic structure below critical T_{cone} has been observed in Sc-doped SrM single crystals (at 270 K for x=1.6) [3]. For Sc-doped SrM nanostructures we reported T_{cone} at temperatures of ~100 K lower [4]. Local maxima in *M*(*T*) dependences in ZFC/FC experiments have been observed for the first time in $Sr_{0.95}Nd_{0.05}Fe_{12-x}Al_xO_{19}$ nanostructures and found considerably lower than those for Sc-doped SrM.

In summary: $Sr_{0.95}Nd_{0.05}Fe_{12-x}Sc_xO_{19}$ soften with increasing doping level *x*, following the dependence of mean grain size on *x*. $Sr_{0.95}Nd_{0.05}Fe_{12-x}Al_xO_{19}$ exhibit low coercivity of ~0.15 T independent of *x*. The hexagonal 10-20 nm thick crystallites may contain only ~4-9 unit cells along their thickness. At the highest doping level x=1.08 two Al ions per one unit cell substitute Fe^{3+}, in $4f_2$ 12k, 2a and $4f_1$ positions. In the case of x=0.36 two ferric ions per 3 unit cells are substituted by Al.

References

[1] Y. Tokura et al. Rep. Prog. Phys. 77(2014)076501
[2] S-P. Shen et al. Phys. Rev. B 90(2014)180404R
[3] Y. Tokunaga et al. Phys. Rev. Lett. 105(2010)257201
[4] A. Hilczer et al. Ceram. Inter. 45(2019)1189

In situ/in operando studies

P63

Investigation of soluto-capillary convection in Ge_xSi_{1-x} melts

J. P. Wöhrle[1], T. Jauß[1], T. Sorgenfrei[1]
[1]Albert-Ludwigs-University Freiburg, Crystallography, Freiburg/ Breisgau, Germany

To solve technical obstacles of industrial semiconductor materials, a profound comprehension of the respective material systems is required. Especially in mixed systems, additional challenges concerning the homogeneity of the grown material are rising due to varying properties of the different chemical composition of the components. Ge_xSi_{1-x} is a semiconductor material which is extremely attractive for various applications like thermoelectrica or microelectronics due to band gap tuning based on composition. A major challenge in growing high quality Ge_xSi_{1-x} bulk crystals from melt, is the fact that the surface tension of Si is 30% higher than that of Ge, but on the other hand the density is only half as large. Due to the large segregation coefficient of Si in Ge of $k_0 \leq 5$, germanium is enriched in front of the solid/liquid interface, what leads to a strong soluto-capillary convection and solutal buoyancy convection. It is shown that this specific type of convection is significantly influencing the shape of the solid-liquid interface and consequently the complete growth process, if a growth technique with free melt surfaces is used. Under normal gravity conditions on earth (1g), it is not possible to investigate the effect of the buoyancy force and different surface-tension-driven convections separately on the crystal growth process. Experiments under µg conditions provide the possibility to investigate the flow behaviour directly on and in the liquid phase without the disturbing impact of the buoyancy convection. Therefore, we use parabolic flight experiments which provide several µg phases up to 22 seconds, which can be used for a quantitative and qualitative determination of the effect of soluto-capillary convection in the Ge-Si system by analysing the movement of small tracer particles on the melt surface during solidification of several Ge_xSi_{1-x} mixtures. A current view on the results of the parabolic flight campaigns of the last years will be presented.

P64

Real-time In-situ Synchrotron Study of Simvastatin Crystallization on Levitated Droplets

Y. Ramisch[1], M. Heilmann[1], C. E. S. Bernardes[2], M. E. Minas da Piedade[2], F. Emmerling[1]
[1]Federal Institute for Materials Research and Testing (BAM) , Structural Analysis, Berlin, Germany
[2]University of Lisbon, Lisbon, Portugal

Simvastatin is one of the most widely used active pharmaceutical ingredients (API) for treatment of hyperlipidemias. The compound is known to exhibit polymorphism,[1,2] and also to exist in stable amorphous phases.[3,4] Controlling its crystallization from solution, therefore, becomes an important issue, to ensure the manufacture of an API with highly reproducible pharmaceutical properties (e.g. shelf-life, dissolution rate, bioavailability).

We studied the crystallization of simvastatin in ethyl acetate, acetone, and ethanol, which differ in polarity and protic character. The studies were carried out by solvent evaporation using acoustically levitated droplets in combination with synchrotron X-ray diffraction, Raman spectroscopy, and imaging analysis. This multi-technique approach has proved to be a very powerful tool to follow the whole crystallization process in real-time and in-situ, and to identify intermediates preceding the final product.[5] Significantly solvent dependent behavior was observed: (*i*) in ethyl acetate, after solvent evaporation, a glassy material was formed, which crystallized on storage over a two-week period; (*ii*) in ethanol a gel was obtained; (*iii*) in acetone the crystallization of stable polymorph I proceeds via an amorphous phase.

[1] Hušák, M.; Kratochvíl, B.; Jegorov, A.; Brus, J.; Maixner, J.; Rohlíček, J. *Struct Chem* 2010, *21*, 511.
[2] Simões, R. G.; Bernardes, C. E. S.; Diogo, H. P.; Agapito, F.; Minas da Piedade, M. E. *Mol. Pharmaceut.* 2013, *10*, 2713.
[3] Graeser, K. A.; Strachan, C. J.; Patterson, J. E.; Gordon, K. C.; Rades, T. *Cryst. Growth Des.* 2008, *8*, 125.
[4] Nunes, T. G.; Viciosa, M. T.; Correia, N. T.; Danede, F.; Nunes, R. G.; Diogo, H. P. *Mol. Pharmaceut.* 2014, *11*, 727.
[5] Thi, Y. N.; Rademann, K.; Emmerling, F. *CrystEngComm* 2015, *17*, 9029.

P65

Determination of proton transfer trajectory in proton conductors by in situ X-ray powder diffraction measurement.

J. Plocek[1], D. Havlíček[2]
[1]Institute of Inorganic Chemistry of the Czech Academy of Sciences, Department of Materials Chemistry, Řež, Czech Republic
[2]Charles University Prague, Department of Inorganic Chemistry, Faculty of Science, Prague, Czech Republic

The study of proton conductivity is important in many different fields and disciplines. The large family of proton conductors includes also salts of oxoacids, which have been studied in our department for several years. The conductivities, whilst not high, were still several orders of magnitude higher than for insulators. We have prepared and described these new compounds, studied their structural properties by single crystal X-ray and neutron diffraction and measured proton conductivity. We were able to define the direction of conductivity in these crystals.

Proton transfer could occur via either the vehicle mechanism or the Grotthus mechanism. In anhydrous salts, the lack of water molecules excludes application of vehicle mechanism. But the application of Grotthus mechanism should cause disorders of the ions through which the protons are transferred at the conditions of proton conductivity.

On the last DGK meeting in Leipzig we have described the changes of X-ray diffraction pattern of CsHSO4 powder sample measured on the "conditions of proton conductivity", i.e. under high voltage (DC). On approx. 60 % diffraction lines we have observed broadening and in some cases also the significant shift of diffraction lines both to higher and smaller angles. The same measurement on normal salt (Cs2SO4 – without hydrogens) did not show any changes of the pattern under 3000 V DC. The analysis of such patterns can show, which lines are affected by proton transfer and define the "direction" of conductivity in the crystal, even the measurement is made on powder sample.

We will present enhanced analysis and new 3D graphic representation of proton transport channels obtained from in situ powder-XRD measurement under high voltage.

Keywords: proton conductivity, powder samples, in situ XRD measurement

P66

***In situ* synchrotron powder diffraction-investigation of the mechanochemical synthesis and phase transition of ZnS**

H. Petersen[1], P. Losch[1], S. Reichle[1], T. Rathmann[1], J. Tseng[2], W. Schmidt[1], M. Etter[3], C. Weidenthaler[1]
[1]Max-Planck-Institut für Kohlenforschung, Heterogene Katalyse, Mülheim, Germany
[2]Shenzhen University, College of Mechatronics and Control Engineering, Shenzhen, China
[3]Deutsches Elektronen Synchrotron (DESY), Hamburg, Germany

In mechanochemical synthesis, the external mechanical impact (chemo-mechanochemical effect) induces reactions. Especially the possibility of avoiding solvents by direct mechanical energy transfer results in an energy-efficient, environmentally friendly and fast synthesis routine, causing growing interest in mechanochemistry. Nevertheless, the reaction mechanisms are not yet fully understood, here *in situ* investigations of model systems can give insights.

The chosen model compound zinc sulfide (ZnS), which can be mechanically synthesized directly from its elements, via an amorphization step [1]. Temperature-dependent (TD) X-ray diffraction (XRD) studies reveal a phase transition of ZnS from cubic sphalerite () to hexagonal wurtzite () at 1003 K [2]. At 1123 K wurtzite transforms back to sphalerite with low reaction rates [3]. Under mechanochemical activation, the transition of sphalerite to wurtzite is not observed, but the reverse transformation from wurtzite to sphalerite occurs [3].

In order to gain some insights in the mechanochemical synthesis as well as phase transitions in the sphalerite-wurtzite-system, *in situ* investigations especially with a good time resolution are mandatory. Therefore, *in situ* synchrotron powder diffraction experiments at PetraIII (P02.1) were performed. A stainless steel milling jar equipped with an X-ray-transparent window was designed for this purpose. The variation of milling frequencies revealed a fast reaction to wurtzite, which is independent from the milling frequency; the reaction acquires an activation energy. After the inition step, the released energy of the exothermic reaction causes a fast reaction. Longer milling times induce the phase transition from the initial wurtzite to the sphalerite structure, which show comparably slow kinetics.

[1]: P. Baláž et al., *Solid State Ionics* 101-103 (1997) 45-51.
[2]: J. Baars et al., *J. Phys. Chem. Solids* 34 (1973) 905-909.
[3]: K. Imamura et al., *Mat. Res. Bull.* 19 (1984) 59-65.

P67

On the oxidation behavior of hexagonal $Y_{0.95}Pr_{0.05}MnO_{3+\delta}$ – application of *in situ* X-ray diffractometry in determination of oxygen incorporation mechanism

K. Cichy[1], K. Jarosz[1], K. Świerczek[1,2], A. Klimkowicz[3], B. Dabrowski[4]
[1]AGH University of Science and Technology, Faculty of Energy and Fuels, Department of Hydrogen Energy, Kraków, Poland
[2]AGH University of Science and Technology, AGH Centre of Energy, Kraków, Poland
[3]Shibaura Institute of Technology, SIT Research Laboratories, Tokyo, Japan
[4]Institute of Physics, Polish Academy of Sciences, Warsaw, Poland

Use of temperature swing process for oxygen separation from air gained recently scientific attention. In such a route, materials capable to reversibly incorporate in their structure substantial amount of oxygen, so called oxygen storage materials (OSM), find their application. Hexagonal rare-earth manganites, $RMnO_{3+\delta}$ (R – smaller rare-earths), can work at relatively low temperatures, ca. 200-400 °C, and this unique feature is related to a fact that incorporated oxygen occupies interstitial positions, instead of the vacancy sites, as in the majority of other OSMs. Since hopping of the interstitial oxygen ions is often characterized by low values of the activation energy, effective transport can take place at lowered temperatures, however, a significant drawback of $RMnO_{3+\delta}$ oxides is their poor ability to operate in air.

In this work, we present how the air-performance of $RMnO_{3+\delta}$ was improved by a proper design of the chemical composition. The main objective of the studies was to explore the nature of oxidation mechanism during operation.

Materials were prepared by a sol-gel method. Crystal structure was studied by means of X-ray diffractometry (XRD). Thermogravimetry was used to evaluate operating temperatures. Complementary *in situ* XRD studies of oxidation kinetics were done for $Y_{0.95}Pr_{0.05}MnO_{3+\delta}$ in the temperature swing process.

It is found that usage of the optimally designed compound could allow to produce 140 m^3-O2/day from 1 ton of the material, over 5 times more than the best value reported so far [1]. Analysis of phase transformation observed by *in situ* XRD led to a conclusion that the limiting factor of the oxidation can be related to the phase boundary influence.

Acknowledgements
This work was funded by National Science Centre Poland, Grant Number 2018/31/N/ST5/02280

References

[1] A. Klimkowicz, K. Cichy, O. Chmaissem, B. Dabrowski, B. Poudel, K. Świerczek, K. Taddei, A. Takasaki, J. Mater. Chem. A, 7 (2018) 2608-2618

P68

***In situ* crystallization and structural studies of allylamine and alcohol co-crystals.**

B. Prus[1,2], Ů Dobrzycki[1], M. Cyrański[1], R. Boese[1]
[1]University of Warsaw, Department of Chemistry, Warsaw, Poland
[2]University of Warsaw, Department of Chemistry, Warsaw, Poland

Alcohols and amines can be considered as excellent cocrystal forming agents, due to the compatibility of intermolecular interactions where both compounds can act as hydrogen bond donor and/or acceptor. In such structures different energy-efficient structural motifs as isolated oligomers(0D), ribbons(1D), layers(2D), etc. can be expected.

The research aimed to investigate the cocrystallization of allylamine with selected aliphatic alcohols (with up to eight carbon atoms), analyze structural motifs and check the dynamic of the molecules in the obtained systems.

The examined mixtures are liquid at ambient conditions, therefore, a laser-assisted *in situ* crystallization method perform directly on the goniometer of the single crystal diffractometer was used[1]. The X-Ray measurement was complement by Raman spectroscopy and DFT calculation.

Among obtained cocrystals, those with methanol, ethanol, and 1-propanol contain molecules arranged in layers with L4(4)8(8) motif[2] of hydrogen bonds. In systems with other alcohols formation of 1D ribbons of hydrogen-bonded molecules is observed. In cocrystal with tert-butanol occur the change of the architecture, which can be attributed to the larger size of the aliphatic group of the alcohol acting as a steric hindrance. The discrepancy between the size of cocrystal components leads to the order-disorder phase transition at lower temperatures with the fragmentation of hydrogen bond chains, turning crystal architecture to 0D.

Relatively small variability of the structural motifs in obtained cocrystals can be used to design new systems with an appropriate arrangement of molecules.

The research was supported by the National Science Center in Poland (Grant SONATA BIS 6 NCN, 2016/22/E/ST4/00461).

Project implemented under the Operational Program Knowledge Education Development 2014-2020 co-financed by the European Social Fund.

[1] R. Boese, Z. Kristallogr., 2014, 229, 595-601
[2] L. Infantes, S. Motherwell, CrystEngComm, 2002, 4, 454–461

Bio-Crystallography II: Enzymes

P69

Optimizing Drug Design by Inhibiting the Steroid Biosynthesis Enzyme CYP51

H. Mohamed[1], S. Bell[1]
[1]University of Adelaide, Chemistry, Adelaide, Australia

Introduction

Tuberculosis (TB) is a common infectious disease triggered by the bacterium, *Mycobacterium tuberculosis* (1). Cytochrome P450 14α-demethylases (CYP51) are key enzymes of sterol biosynthesis in eukaryotes, catalysing the removal of the 14α-methyl group from sterol precursors but their role in bacteria is less clear (2). **The aim** of this study is to understand substrate and inhibitor binding to the bacterial cytochrome P450 CYP51 enzymes from *M. tuberculosis* (MtbCYP51) and *M. marinum* (MmarCYP51). A combination of computational techniques and virtual screening tools will be used to find new compounds that complement the substrate binding pocket of MtbCYP51.

Methods

A structure-based pharmacophore was generated and validated using the crystal structures of enzyme-inhibitor complexes of MtbCYP51. Virtual screening was carried out on the ZINC database (3) using the validated pharmacophore model and various docking algorithms with MtbCYP51. Substrate binding studies were performed with different ligands with MtbCYP51 and MmarCYP51.

Results

Comparison of the chemical structures from the virtual screening highlighted the high prevalence of a pyrimidine scaffold. Molecular docking showed that the pyrimidine moiety was involved in several binding interactions with residues in the MtbCYP51 active site. Azole containing compounds were also found in the virtual screen and are also present in many reported MtbCYP51 inhibitors (2).

Conclusion

Several inhibitors with different substrate binding modes were identified for each of the CYP51 enzymes. This could lead the development of targeted-drug design for the diseases caused by different species of Mycobacterium.

References

[1] Organization WH. Geneva: WHO. 2005.
[2] Podust LM *et al*. Proceedings of the National Academy of Sciences. 2001;98(6):3068-73.
[3] Irwin JJ *et al*. Journal of chemical information and modeling. 2005;45(1):177-82.

Figure 1

Figure 2

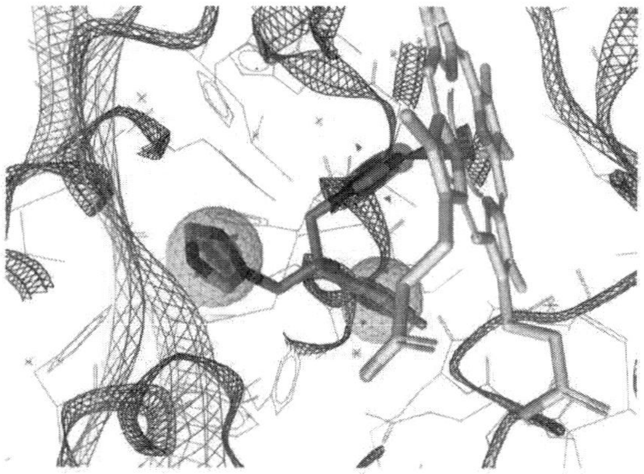

P70

Dual-function tail tubular proteins of bacteriophages

A. Pyra[1], K. Filik[2], E. Brzozowska[2]
[1]Wrocław University of Science and Technology, Faculty of Chemistry, Wrocław, Poland
[2]Hirszfeld Institute of Polish Academy of Sciences, Biology, Wrocław, Poland

Both, tail tubular proteins A (TTPAs) and tail tubular proteins B (TTPBs) have been considered as structural proteins of environmental bacteriophages. But recent studies revealed that some of TTPs are dual-function proteins due to their hydrolytic activities [1-3].

We have researched TTPA and TTPB of *Klebsiella phage* KP32, and TTPAs of *Klebsiella* phage KP34 and *Yersinia* phage phiYeO3-12.

All TTPs mentioned above were overexpressed in *E.coli* BL21(DE3)pLysS cells and purified by nickel-affinity and gel-filtration chromatography. Bacterial exopolysaccharides (EPSs) or chromogenic substrates were used to indicate and assess the enzymatic activity of TTPs. Both antibiofilm and agar overlay tests were carried out to detect antibacterial properties of TTPs. All of these proteins were also tried to be crystallized to determine their 3D structures.

The hydrolytic activity of TTPAs was determined towards maltose and trehalose and TTPB displays activity towards bacterial EPSs. TTPs of *Klebsiella* phages showed activity against biofilm-forming bacterial strains causing chronic, life-threatening infections. The crystals of TTPs of *Klebsiella* phages were obtained but only the crystal structure of TTPA of phage KP32 was determined so far - the protein molecules adopt a novel tetrameric structure with α-helical domains on one side and β-strands and loops on the other [1]. The crystal structures of the other TTPs are still being studied.

[1] A. Pyra, E. Brzozowska, K. Pawlik, A. Gamian, M. Dauter, Z. Dauter, *Scient. Rep.*, 7 (2017) 2223.
[2] E. Brzozowska, A. Pyra, K. Pawlik, M. Janik, S. Górska, N. Urbańska, Z. Drulis-Kawa, A. Gamian, *Scient. Rep.,* 7 (2017) 18048.
[3] E. Brzozowska, A. Pyra, K. Pawlik, S. Górska, A. Gamian, *Preprints*, doi:10.20944/preprints201803.0075.v2

This research was supported by National Science Center (Poland), grant no.2017/26/E/NZ1/00249.

Organic molecules and coordination compounds

P71

Porous Mixed Metal Organic Frameworks as Precursors for Heterogeneous Catalysts

H. Gildenast[1], U. Englert[1]
[1]RWTH Aachen University, Institute of Inorganic Chemistry, Aachen, Germany

In the pursuit of counteracting climate change new approaches are needed to prevent greenhouse gas emissions. The media focus mainly on the reduction of CO_2 to achieve this goal, but several other gases such as N_2O have a far more severe greenhouse effect and need to be taken into consideration.[1] The decomposition of N_2O can be catalyzed by Ag but only at high temperatures. To make this process greener a different approach is needed. In 2016 we reported a highly active catalyst at moderate temperature for this reaction with Ag nanoparticles in porous Yb_2O_3, a material that is obtained after thermal decomposition of a mixed metal organic framework (MMOF).[2] The costly synthesis of the organic linker, however, rendered this process irrelevant for economy.
Here we report the construction of a semi-porous MMOF with approximate **fee** topology constructed by an inexpensive ligand. The network contains stoichiometric amounts of Ag^+ ions and one of five different M^{3+} cations (Al, Cr, Fe, Co, Ga). The 3D network was found to be structurally stable towards the removal of one of the solvent molecules. The removal could be observed with consecutive SCXRD measurements showing a decline of the occupation of the solvent molecule. The M^{3+} site can also be occupied statistically by two and possibly more of those metal cations. Also, the network can be thermally decomposed in a vigorous reaction. The remains were analyzed by PXRD and show promising results for future application.

Figure 1: Schematic structure of the MMOF, as structural formula (left) and a simplified GTECS3D[3] analysis (right).

[1] F. Kapteijn, J. Rodriguez-Mirasol, J. A.Moulijn, *Appl. Catal. B*, 1996, *9*, 25–64.
[2] M. Konkol, M. Kondracka, P. Kowalik, W. Próchniak, K. Michalska, A. Schwedt, C. Merkens, U. Englert, *Appl. Catal. B*, 2016, *190*, 85–92.
[3] K. Lamberts, C. Merkens, R. Wang, U. Englert, D. Hons, S. Grüter, Y. Guo, S. Porsche, A. Hamacher, D. Bündgens,T. Kuhlen, *Z. Kristallogr.* 2012, *Suppl. 32*, 117.

Figure 1

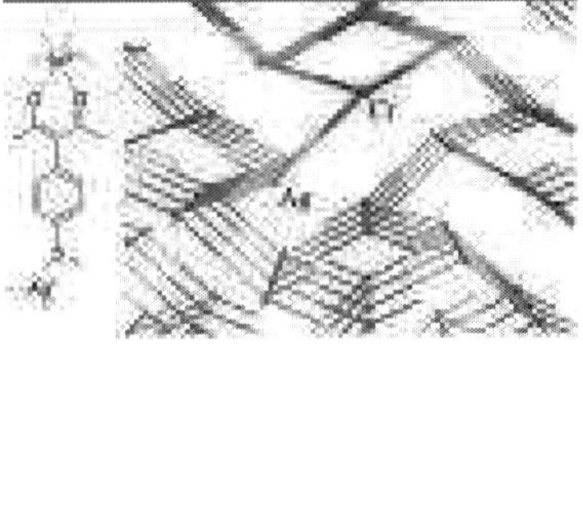

P72

From Nano-Balls to Nano-Bowls: Pentaphosphaferrocene-Based Supramolecule with Open Architecture

E. Peresypkina[1], A. Virovets[1,2], M. Scheer[1]
[1]University of Regensburg, Regensburg, Germany
[2]Novosibirsk State University, Novosibirsk, Russian Federation

Since 2003, we have been using pentaphosphaferrocenes $[Cp^R\text{-}Fe(\eta_5\text{-}P_5)]$ ($Cp^R = \eta^5\text{-}C_5R_5$, R = Me, CH_2Ph) as building blocks to obtain giant supramolecules in the reaction with Cu(I) halides (CuX, X = Cl, Br [1-3]). The most symmetrical supramolecule ideally has 80 Cu and P atoms in the hollow spherical $\{CuX\}_{20}(cyclo\text{-}P_5)_{12}$ metallophosphorus core (so called 80-vertex core, Fig.A), which is able to encapsulate guest molecules like Cp_2Fe or $o\text{-}C_2B_{10}H_{12}$, demonstrating guest-driven template effect.

Recently we have found that in the presence of larger triple-decker complexes $[(CpCr)_2(\eta^5\text{-}As_5)]$ and $[(CpMo)_2(\eta^5\text{-}P_4S)]$ the supramolecules with open architecture can be obtained. The bowl-like 70-vertex $\{CuX\}_{15}(cyclo\text{-}P_5)_{11}$ core can be derived from 80-vertex one by elimination of one ($cyclo$-P_5) and 5 CuX units (Fig.B) [4]. Some of Cu and X positions can be statistically absent, as in 80-vertex cores [1-3]. The guest molecule protrudes in the opening in the host core.

Fast crystallization results in the centrosymmetric $R\bar{3}$ phase, where the supramolecules are disordered, mimicking "80-vertex supramolecules". Only closer look at the atomic displacement parameters and residual electron density allowed us to discover the presence of the open architectures. Slower crystallization resulted in the ordered (sp. gr. *Cc* or *Pc*, racemic twining) phases. Orientationally ordered supramolecules form head-to-tail chains along *b* axis supported by $P_5...C_5$ and $P_5...C_5$ π-stacking interactions (Fig.C).

[1] E. Peresypkina, C. Heindl, A. Virovets, M. Scheer (2016) *Structure and Bonding* (2016) 174, 321
[2] A. Schindler, C. Heindl, G. Balázs, et al (2012) *Chem. A Eur. J.* 21, 6208
[3] F. Dielmann, M. Fleischmann, C. Heindl, et al (2015) *Chem. A Eur. J.* 18, 829
[4] H. Brake, E. Peresypkina, C. Heindl, et al (2019) *Chem. Sci.* 10, 2940

Figure 1

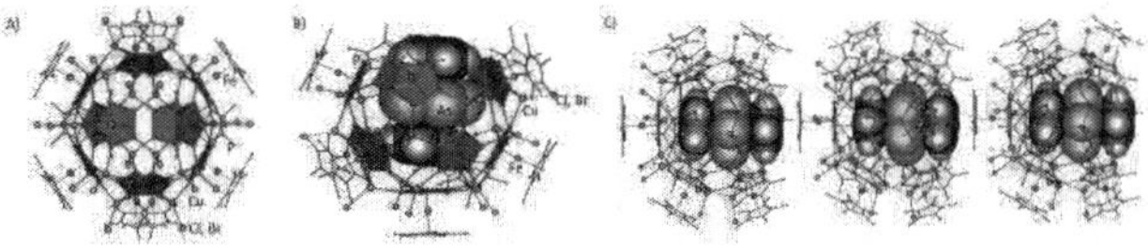

P73

Characterization and thermal behaviour of the iron dietary supplement ferrous glycine sulphate pentahydrate

G. Gallo[1,2], S. Bette[1], R. E. Dinnebier[1]
[1]Max-Planck-Institute for Solid State Research, Stuttgart, Germany
[2]University of Salerno, Department of Chemistry and Biology "A. Zambelli", Salerno, Italy

Introduction

Complex metal compounds of glycine (gly) are of particular interest as source of trace elements and amino acids and used as active pharmaceutical ingredient (API) in different formulations for dietary supplements. $Fe(gly)SO_4\cdot 5H_2O$(**1**), a therapeutic agent which treats iron deficiency anaemia in infants and women, was characterized with different techniques (XRPD, SEM, IR). Thermal studies (TGA, DTA and variable temperature *in situ* X-ray powder diffraction (VT-XRPD) measurements revealed the phase transition from **1** to the anhydrous form through Fe(gly)SO4·3H2O(**2**) (Fig. 1).

Objectives

The study of **1** using XRPD aims to gain knowledge about crystal structure which plays an important role in the process of finding new efficient pharmaceutical drugs, API-quantification for quality control and of detecting drug fraud and abuse.

Results

The crystal structures of the two hydrate forms were determined and fully described. Compound **2** exhibits, compared to **1**, a different coordination environment of the iron sites where glycine ligands bridge iron forming a 1D polymeric chain structure. *In situ* VT-XRPD confirmed the phase transition from **1** towards the anhydrous from through **2**. From detailed structural comparison the mechanism of the phase transitions can be concluded.

Conclusions

This study gives a global understanding of one of the most important API used as iron dietary supplement and provides a reliable basis for the substance identification by XRPD and IR spectroscopy. The investigated thermal behaviour of the API can be used to derive suitable storage and drying conditions.

Fig. 1. Representation of the transition from **1** (left) to its anhydrous form (right) through **2** (middle). The directions of the movements are highlighted in black for sulphate ions and in light blue for glycines. The released water molecules in **1** are depicted in light blue.

Figure 1

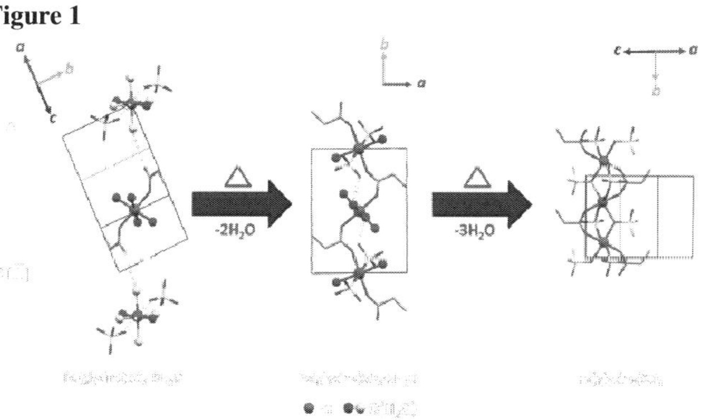

P74

Studies of Fe complexes of N_2O_2 bis-phenolate ligand based on ethylenediamine

K. Kałduńska[1], E. Safaei[2], A. Kozakiewicz[1], A. Wojtczak[1]
[1]Nicolaus Copernicus University, Faculty of Chemistry, Toruń, Poland
[2]Institute for Advanced Studies in Basic Sciences (IASBS), Zanjan, Iran

Introduction

Complexes of bisphenolate ligands with Fe/Mn/Cu were synthesized as analogues of the centre of catechole dioxygenase. [1] An analogous structure is often correlated with similarity in the catalytic mechanism.

Objectives

The aim was to determine the structure of Fe(II/III) complexes with a ligand (**L**) (**Fig.1.**)

Materials and methods

The measurements were performed on an Oxford Sapphire CCD diffractometer, T = 293K, MoKα (l = 0.71069 Å), with CrysAlis CCD/Red package. [2] The structures were solved and refined with SHELX-2017 package. [3]

Results

Structural studies were performed for the ligand **L**, and complexes **[FeLQ]** and **[(FeL)$_2$O]**. In the **[FeLQ]** complex, the central atoms have a deformed octahedral coordination sphere formed by **L** and 5-chloro-7-iodoquinolin-8-ol (**Q**). In the dimer **[(FeL)$_2$O]**, iron atoms are bridged by axial oxo O^{2-}, and have a square pyramid environment N_2O_3. In the free ligand, the rings are parallel to each other, the dihedral angle is 4.21°. In **[(FeL)$_2$O]**, the similar dihedral angle of 7.08° is found, and the displacement of Fe from the N_2O_2 plane is 0.635Å. In contrast, in **[FeLQ]** the dihedral angle is 82.6°. The change in the spatial arrangement of the ligand **L** is related to the difference between the detected R,S and R,R configuration.

Conclusion

Based on the analysis of biomimetic structures and previous research, the catalytic activity of **[(FeL)$_2$O]** dimer is expected to be greater than that of **[FeLQ]**, which is associated with better access to the reaction centre. Tests the catalytic activity are in progress.

References

[1] E. Safaei, A. Wojtczak, N. Naghdi, Z. Jagličić, Polyhedron, 2016, 109, 190-198.
[2] CrysAlisPro 1.171.38.43 package of programs, Rigaku OD, 2015.
[3] G. M. Sheldrick, Acta Cryst., 2015, C71, 3-8.

Figure 1

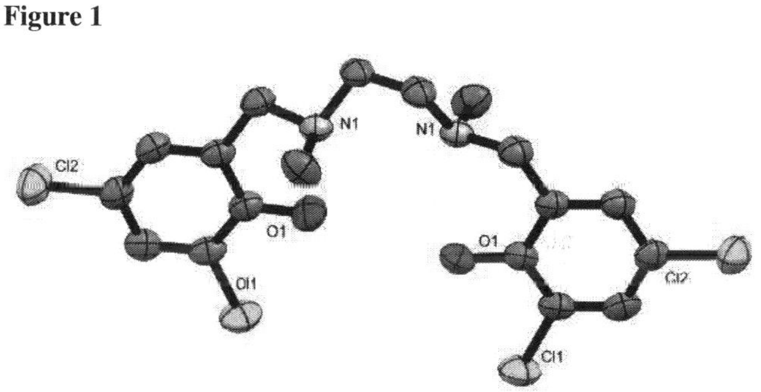

P75

Solvothermall synthesis and characterization of 1D Cu(II) coordination polymer with dianion of pirydyne-2,3-dicarboxylate acid

A. Kochel[1], K. Twaróg[2], M. Hołyńska[3]
[1]Wrocław University of Science and Technology, Faculty of Chemistry, Wrocław, Poland
[2]Wrocław University of Science and Technology, Faculty of Chemistry, Wrocław, Poland
[3]Philipps-Universität Marburg, Fachbereich Chemie and Wissenschaftliches Zentrum für Materialwissenschaften, Marburg, Germany

Coordination polymers (CPs) are functional materials due to such properties as luminescence, conductivity, catalytic effect, sensing effect, ion exchange and magnetism [1].

Coordination polymer $[Cu(2,3\text{-pdc})H_2O]_n$ was obtained under solvothermal conditions in pressure reactor Berghoff BF100. The obtained compound crystallizes in triclinic crystal system, space group , a = 7.434(3) Å, b = 7.523(4) Å, c = 7.881(3) Å, α = 62.68(5)°, β = 179.02(5)°, γ = 78.90(5)°, V = 381.5(3) Å^3, Z = 2. It was characterized via EA, IR, EPR, X-ray diffraction studies, magnetometry and fluorescence spectroscopy.

The coordination environment of the central Cu2+ ion adopts shape of distorted tetragonal pyramid. The polymer forms a 1D chain along [100] and is also involved in intermolecular hydrogen bonds. Magnetic susceptibility data and their modelling suggest the presence of antiferromagnetic interactions with Néel temperature at 6.2 K.

Fig. 1: 1D coordination polymer $[Cu(2,3\text{-pdc})H_2O]_n$.

[1] Kitagawa S., Kitaura R., Noro S, *Angew. Chem., Int. Ed.* 2004, 43, 2334.

Figure 1

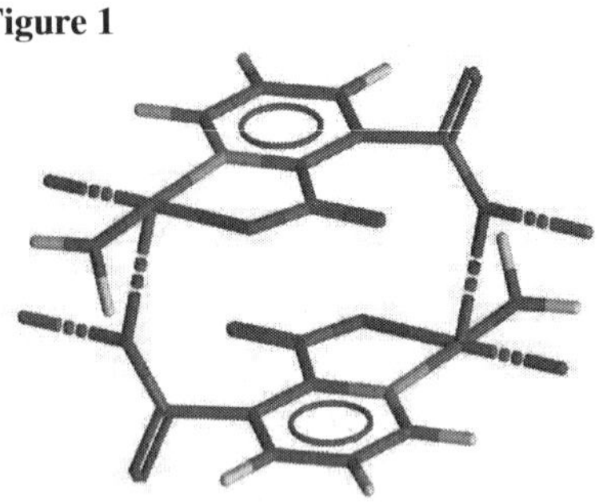

P76

Crystal structure, theoretic and spectroscopic analysis of three salts of 2m3na with inorganic acids

V. Medviediev[1], M. Daszkiewicz[1]
[1]Institute of Low Temperature and Structure Research, Polish Academy of Sciences, Division of Structure Research, Wrocław, Poland

The 2-methyl-3-nitroaniline (2m3na) is an isomer of 2-methyl-4-nitroaniline which possesses nonlinear optical (NLO) properties. The amino group in nitroanilines can be protonated and therefore their salts can be easily obtained under acidic conditions. So, a search of compounds with NLO properties can be expanded for a large group of organic ionic compounds.

Previously, some complexes of 2-methyl-4-nitroaniline and 2-methyl-5-nitroaniline with inorganic acids were studied [1,2]. As a continuation of those studies, here we present crystal structures of three new compounds (H2m3na)Cl·H2O (**1**), (H2m3na)Br (**2**) and (H2m3na)I (**3**).

Crystal structures of obtained salts were determined and their temperature stability were checked in the range of 295–100 K using XRD. The FT-IR spectra were measured at room temperature. Theoretical DFT studies of the relaxed potential energy surfaces were carried out for rotation of all the residual groups of the H2m3na+ ion.

All the compounds crystallize in centrosymmetric space groups: *P*21/*c* for **1** and **2**, and *Pbcm* for **3**. Analysis of hydrogen bonding patterns were performed using a combination of XRD, IR and computational data. The calculations for the H2m3na+ ion showed that global energy minimum corresponds to the conformation of the ion when the nitro group is rotated by approximately 40 deg with respect to benzene ring. This result agrees with XRD data for all presented structures.

No phase transitions were found in the range of 295-100 K. Asymmetric part of the unit cell of the chloride salt contains additional water molecule. The strongest hydrogen bonds exist in **2**, and the weakest ones in **1**. Surprisingly, calculated rotational barrier for the nitro group is less than 2.0 kcal/mol. This value is much lower than previously obtained for other isomers of H2m3na+ ion [2].

References

[1] M. Daszkiewicz, Cryst. Growth Des., 2013, 13, 2277–2285.
[2] V. Medviediev, M. Daszkiewicz, *Acta Cryst.* B**75**, 2019, doi.org/10.1107/S2052520619012472

P77

The influence of the conformation of the bridging ligands on the spin crossover transitions in $[Fe(bbtre)_3](ClO_4)_2{\cdot}2CH_3CN$

M. Książek[1], M. Weselski[2], A. Dreczko[2], V. Maliuzhenko[2], M. Kaźmierczak[2], A. Tołoczko[2], J. Kusz[1], R. Bronisz[2]
[1]University of Silesia in Katowice, Faculty of Science and Technology, Institute of Physics, Chorzów, Poland
[2]Wrocław University of Science and Technology, Faculty of Chemistry, Wrocław, Poland

Introduction

Recently two coordination polymers $[Fe(ebbtr)_2(CH_3CN)_2]X_2{\cdot}4CH_3CN$ (ebbtr=1,4-di(5-ethyl-1,2,3-triazol-1-yl)butane, X=ClO_4^-,[1] $CF_3SO_3^-$[2]) have been synthesized. They exhibit unusual properties resulting from interplay between change of spin state and structural alterations.

Objectives

An occurrence of uncommon spin crossover properties, associated with presence of the flexible building blocks, was an encouragement to study another ligands based on 1,5-disubstituted-1,2,3-triazoles with butylene spacer but joined through carbon atoms (bbtre).

Materials and methods

Temperature dependent measurements of the magnetic susceptibility were carried out with a SQUID magnetometer. Structural studies were performed with a four-circle SuperNova X-ray diffractometer.

Results

Slow cooling of $[Fe(bbtre)_3](ClO_4)_2{\cdot}2CH_3CN$ results in high temperature (HT)↔low temperature (LT) structural phase transition and incomplete LT(HS)↔(HS/LS) spin crossover (SCO) [3]. Rapid cooling leads to complete SCO (HT(HS)→HT1(LS)). Heating triggers reversible HT1(HS)→HT1(LS) SCO. Above 170K there is a structural phase transition HT1(HS)→HT2(HS) leading to formation of modulated structure. It is also possible to carry out HT2(HS)↔(LS) SCO. Heating above 200K results in formation of LT(HS) structure and LT(HS)↔(HS/LS) SCO can be accomplished.

Conclusion

Conformational changes of bridging ligands are closely related with unusual spin crossover properties depending on occurrence of three spin crossover transitions arranged in two ways.

Acknowledgements

This work was supported by the Polish National Sciences Centre Grant No. DEC-2014/15/B/ST5/04771.

References
[1] M. Weselski, M. Książek, D. Rokosz, A. Dreczko, J. Kusz, R. Bronisz, *Chem. Commun.*, 2018, 54, 3895-3898.
[2] M. Weselski, M. Książek, P. Mess, J. Kusz, J. R. Bronisz, *Chem. Commun.*, 2019, 55, 7033-7036.
[3] M. Książek, M. Weselski, A. Dreczko, V. Maliuzhenko, M. Kaźmierczak, A. Tołoczko, J. Kusz, R. Bronisz, *in review*

P78

New structural parameter for a description of the geometry of four-coordinate compounds

A. Okuniewski[1], D. Rosiak[1]
[1]Gdansk University of Technology, Department of Inorganic Chemistry, Gdańsk, Poland

Structural parameters (τ) are the numbers that range from 0 to 1 and describe the geometry of the coordination center. The first such parameter was developed in 1984 by Addison and coworkers for five-coordinate compounds (τ_5). [1]

Analogic parameter for four-coordinate compounds (τ_4) was invented in 2007 by Yang *et al.* [2] It shows if the geometry is similar to square planar ($\tau_4 = 0$), tetrahedral ($\tau_4 = 1$) or somewhere in between (Fig. 1). The formula is:

$$\tau_4 = (360° - (\alpha + \beta)) / (360° - 2\theta)$$

where $\beta > \alpha$ are the greatest valence angles at the coordination center, $\theta \approx 109.47°$ is the ideal tetrahedral angle.

Unfortunately, τ_4 parameter does not distinguish α and β angles, so structures of significantly different geometries can have similar τ_4 values. To overcome this issue we were about to develop a new parameter τ_4' that adopts values similar to τ_4 but better differentiates the examined structures.

As a result, we have proposed the following formulae (variables are the same as above): [3]

$$\tau_4' = (\beta - \alpha) / (360° - \theta) + (180° - \beta) / (180° - \theta)$$

Extreme values of both parameters denote exactly the same geometries, however $\tau_4' \leq \tau_4$, so the deviation from ideal tetrahedral geometry is more visible.

Fig. 1. Selected values of τ_4 and τ_4' parameters. For animations see QR-code or [5,6].

The new parameter τ_4' that better indicates a deviation from tetrahedral geometry was proposed. [3,4] Additionally, the on-line script that calculates τ_4, τ_4' and τ_5 was written. [6] Similarities and differences, as well as the correlation of τ_4 and τ_4' parameters, will be discussed in the poster.

References

[1] A. W. Addison, N. T. Rao, J. Reedijk, *et al.*: *J. Chem. Soc. Dalton Trans.* (1984) 1349.
[2] L. Yang, D. R. Powell, R. P. Houser: *Dalton Trans.* (2007) 955.
[3] A. Okuniewski, D. Rosiak, J. Chojnacki, B. Becker: *Polyhedron* 90 (2015) 47.
[4] D. Rosiak, A. Okuniewski, J. Chojnacki: *Polyhedron* 146 (2018) 35.
[5] *en.wikipedia.org/wiki/Geometry_index*
[6] *kchn.pg.gda.pl/geom*

Figure 1

P79

Crystal structure of Cu(II) complex salts with 4*H*-1,2,4-triazole-4-amine derivatives

E. Ganczar[1], A. Białońska[1]
[1]Wrocław University of Science and Technology, Faculty of Chemistry, Wrocław, Poland

4*H*-1,2,4-triazole-4-amine and its derivatives play an important role in supramolecular coordination chemistry.[1] Different possibilities of bridging transition metal ions through an N1, N2 - coordination mode, related to the nature of the substituent at the triazole ring, leads to the diversity of resulted crystal structure from triple core complexes to 1D, 2D and 3D coordination polymers.[2] Besides of ligand topology, the formation of coordination polymers is significantly influenced by the properties of the complexation site, environment (solvent and its presence in the network), kind of an anion, as well as the geometry and nature of the metal core.[3,4]

Preliminary studies conducted in our team have shown that substitution in the orto position of the phenyl ring of the *N*-[(*o*-methyl)benzylidene]-4*H*-1,2,4-triazol-4-amine ligand can result in the formation of a copper(II) complex salt in the form of a 1D coordination polymer with helical topology. Replacement of the hydrogen atom with a fluorine atom in methyl group heavily change the supramolecular structure of this compounds causing the helical chain to unravel.

On the poster, structural studies of the copper(II) complexes compounds with *N*-[(*o*-trifluoromethyl)benzylidene]-4*H*-1,2,4-triazol-4-amine (CF3Phtrz) will be presented. During our investigation on the coordination chemistry of Cu(II) metal with Schiff based on 4H-1,2,4-triazole-4-amine, we have found that anions and type of derivatives of 4*H*-1,2,4-triazole-4-amine play important role in type of 1D linear chain and we show that solvent system gain important role in creation of voids in this type of coordination polymer.

Fig.1 *N*-[(*o*-trifluoromethyl)benzylidene]-4*H*-1,2,4-triazol-4-amine (CF3Phtrz)

References
[1]G. Aromi, L. A. Barrios, O. Roubeau, P. Gamez, *Coord.Chem.Rev.*, 2011, 255, 485.
[2] J. G. Haasnoot, *Coord.Chem.Rev.*, 2000, 131.
[3] J. M. Lehn, *Struct.Bonding* (Berlin), 1973, 16, 1.
[4] Y. Wang, B. Ding, P. Cheng, D. Liao, S. Yan, *Inorg.Chem.*, 2007, 46, 2002.

Figure 1

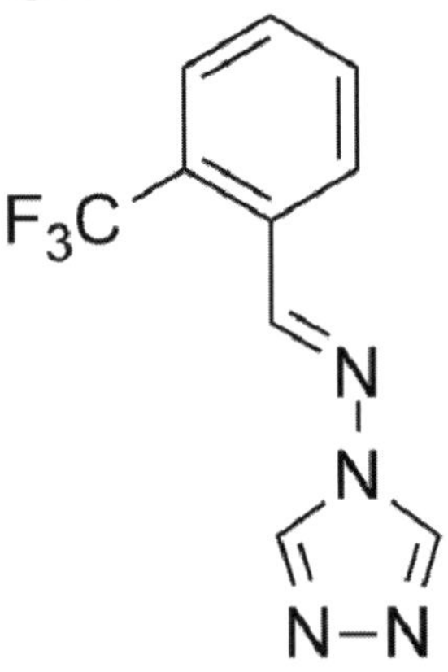

P80

Synthesis, crystal and molecular structure and biological activity of new chiral sulfonamides with pyrazolo[4,3-e][1,2,4]triazine core

Z. Karczmarzyk[1], M. Mojzych[1], Z. Bernat[1]
[1]Siedlce University of Natural Sciences and Humanities, Faculty of Exact and Natural Sciences, Siedlce, Poland

The pyrazolo[4,3-*e*][1,2,4]triazines are the natural compounds with wide antibiotic and antitumour properties and they were found as extracellular metabolites of microorganism of the class Pseudomonas fluorescens var. pseudoiodinine and Nostoc spongiaeforme.1 The combination of the natural pyrazolo[4,3-*e*][1,2,4]triazine compounds with a sulfonamide moiety as pharmacophore group has enabled the design of novel derivatives with potential antibacterial, antimalarial, hypotensive, diuretic, hypoglycemic, antithyroid, antiparasitic, anti-inflammatory and antiglaucomatous properties.2 Therefore, the main subject of our research was focused on the synthesis and investigation of structure–activity relationship of newly synthesized series of sildenafil analogues3 and aniline substituted pyrazolo[4,3-*e*][1,2,4]triazine sulfonamides.4

Continuing our study we have designed, synthesized and evaluated molecular structure of new groups of sulfonamides with potential anticancer activity. One group constitute sulfonamides obtained from 1-phenyl-pyrazolo[4,3-*e*][1,2,4] triazine and the other group contains a tricyclic heterocyclic core with a terminal tetrazole ring that can occur in tautomeric equilibrium with the corresponding 5-azido derivative of pyrazolo [4,3-*e*][1,2,4] triazine.

[1] Smirnov, V. V.; Kiprianova, E. A.; Garagulya, A. D.; Esipov, S. E.; Dovjenko, S. A. FEMS Microbiol. Lett., 1997, 153, 357.
[2] A. Kleemann, J. Engel, B. Kutscher, D. Reichert, Eds; Pharmaceutical Substances, Syntheses, Patents, Applications, (Thieme, Stuttgart, 1999).
[3] Mojzych, M.; Ceruso, M.; Bielawska, A.; Supuran, C. T. Bioorg. Med. Chem., 2015, 23, 3674.
[4] Mojzych, M.; Subertová, V.; Bielawska, A.; Bielawski, K.; et al. Eur. J. Med. Chem., 2014, 78, 217.

Figure 1

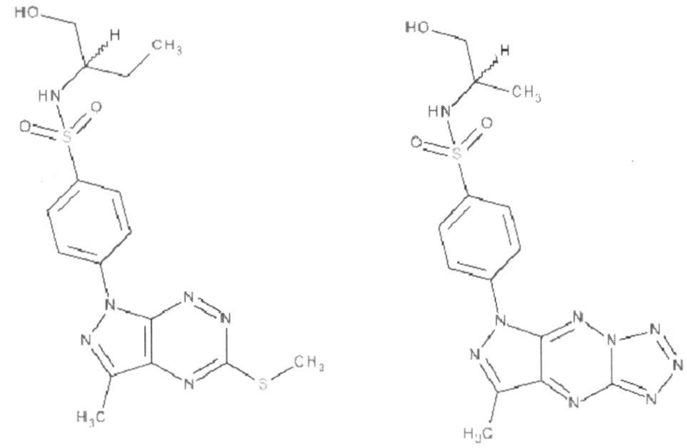

P81

The influence of solvents on the crystal structure of thiourea derivative

D. Rosiak[1], A. Okuniewski[1], J. Chojnacki[1]
[1]Gdansk University of Technology, Department of Inorganic Chemistry, Gdańsk, Poland

One of the main goals of crystal engineering is to learn about intermolecular interactions to design and obtain crystals with a specific structure and desired properties. [1] Despite many studies focused on obtaining specific crystals, predicting their exact structure and properties it is still a difficult task. [2-4] During crystallization, the inclusion of solvent molecules into the crystal lattice may occur, which happens either to minimize voids in the lattice or to compensate for the imbalance of hydrogen bond donors and acceptors in the crystal-forming molecules.

During our research on thioureas, we have performed recrystallization of 1-benzoyl-3-(4-methylpyrimidin-2-yl)thiourea from acetone and have obtained solvated crystals. We decided to check if this compound cocrystallizes with other molecules. For this purpose, the crushed and dried crystals were recrystallized from several standard organic solvents. When dimethyl sulfoxide was used, the compound crystallized without solvent, while in the case of methanol, dichloromethane, and acetone, the compound crystallized as isostructural hemisolvates. In the above three structures, thiourea molecules form a hydrogen-bonded supramolecular, two-dimensional network with solvent molecules placed in the cavities. The third type of structure was obtained using chloroform. The above crystal structure can be described as hydrogen-bonded one-dimensional double ribbons of thiourea that are connected through solvent molecules (employing weak halogen bonds).

Structural and physicochemical details will be presented in the poster.

References

[1] G. R. Desiraju, *J. Am. Chem. Soc.*, 135 (2013) 9952–9967
[2] T. S. Thakur, R. Dubey, G. R. Desiraju, *Annu. Rev. Phys. Chem.*, 66 (2015) 21–42
[3] A. J. Cruz-Cabeza, *Acta Crystallogr. Sect. B-Struct. Sci.Cryst. Eng. Mat.*, 72 (2016) 437–438
[4] J. Nyman, S. M. Reutzel-Edens, *Faraday Discuss.*, 211 (2018) 459–476.

Figure 1

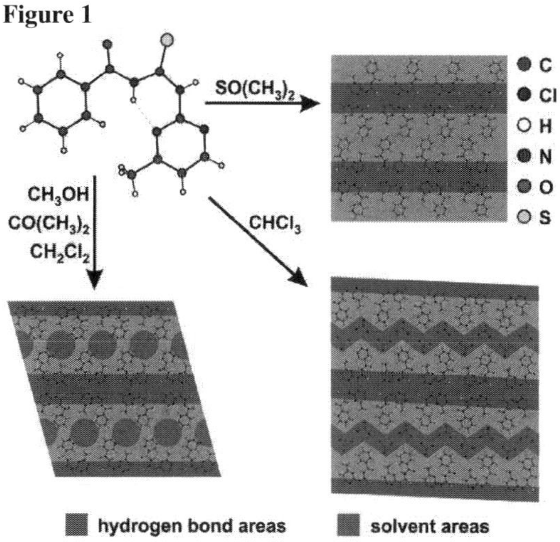

P82

The Structural Study of New Thiosemicarbazones with Potential Antituberculosis Activity

W. Wysocki[1], M. Pitucha[2], Z. Karczmarzyk[1], M. Drózd[2]
[1]Siedlce University of Natural Sciences and Humanities, Faculty of Exact and Natural Sciences, Siedlce, Poland
[2]Medical University of Lublin, Independent Radiopharmacy Unit, Faculty of Pharmacy , Lublin, Poland

Tuberculosis (TB) is one of the most seriuos human diseases and public problems throught the world [1]. A multidrug regimen taken for at least 6 months was established in the 1970s and is still used today. However, the high burden of tuberculosis infections in regions with limited health care resources has led to frequent treatment interruption and subsequent failure resulting in the rise of multi-drug resistant (MDR), and extensively drug resistant (XDR) strains [2]. The development of counterparts with better pharmacokinetic properties and less toxicity appears to be one of the methods of searching for new potential drugs [3].

The aim of this study was to identify a candidate drug for the development of anti-tuberculosis therapy from previously synthesized compounds based on the thiosemicarbazones compounds. Thiosemicarbazones **1–7** were formed by the condensation of an aldehyde or ketone with a appropriate thiosemicarbazide. As a starting materials for synthesis of compounds **1–4** ware used 4-methyl-3-thiosemicarbazide and substituted benzaldehyde, for compounds **5–7** arylthiosemicarbazide with substituted acetophenone.

In the present communication, we will discuss the results from single-crystal X-ray diffraction studies and the theoretical calculations at DFT/B3LYP/6-311++G(d,p) level for **1–7** compounds.

References

[1]. Volynets G.P., Tukalo M.A., Bdzhola V.G., Derkach N.M., Gumeniuk M.I., Tarnavskiy S.S., Starosyla S.A., Yarmoluk S.M. *J Antibiot 72*, 218 (2019).
[2]. Ollinger J., Bailey M.A., Moraski G.C., Casey A., Florio S., Alling T., Miller M.J., Parish T. *PLoS ONE 8(4)*, e60531 (2013).
[3]. Pitucha M., Karczmarzyk Z., Swatko-Ossor M., Wysocki W., Wos M., Chudzik K., Ginalska G., Fruzinski A., *Molecules 24,* 251 (2019).

Figure 1

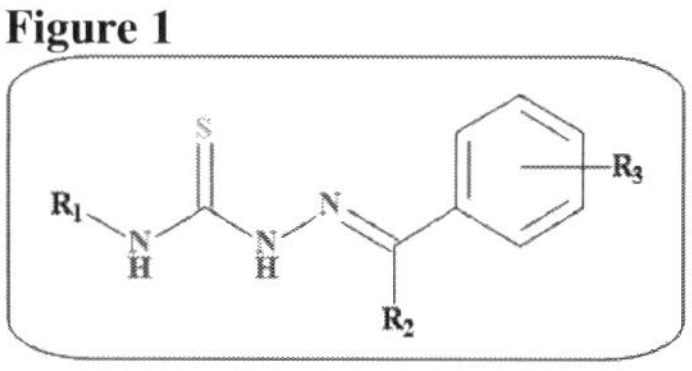

Compound	R_1	R_2	R_3
(1)	Me	H	4-Br
(2)	Me	H	2-Cl
(3)	Me	H	3,4-diCl
(4)	Me	H	3,4-diMeO
(5)	Ph	Me	(-)
(6)	Ph	Me	4-NH_2
(7)	2-ClPh	Me	(-)

P83

Reactivity of [Ru^{II}(terpy)(N^N)Cl]Cl complexes.

A. Kozakiewicz[1], M. Chrzanowska[1], A. Katafias[1]
[1]Nicolaus Copernicus University, Faculty of Chemistry, Department of Biomedical and Polymer Chemistry, Toruń, Poland

Introduction

Ru complexes exhibit a broad range of applications. They are promising antitumor drug candidates, water oxidation catalysts and reported to control the redox biology of cells[1].

Objectives

Since Ru complexes are activated via aquation of a monodentate ligand, our main goal was to tune their substitution behavior using various non-leaving N^N-donor ligands.

Materials and methods

The aqueous behavior of the complexes was studied spectrophotometrically within the visible range using a Shimadzu UV-1601 PC spectrophotometer. The X-ray data were collected with an Oxford Sapphire CCD diffractometer using MoKα radiation and ω-2θ method. The structures were solved and refined with the use of SHELX2014 program packages[2].

Results

Our recent studies focused on a series of [Ru(terpy)(N^N)Cl]Cl complexes, where N^N = 2,2'-bipyridine (bipy), ethylenediamine (en), 2-(aminomethyl)pyridine (ampy),[1,3] N,N,N',N'-tetramethylethylenediamine (tmen) and 1,10-phenantroline (phen) showed that the lability of chloride and the acidity of their aqua derivatives are affected by electronic (σ-donor and π-back bonding) and steric effects provided by bidentate ligands.

Fig. Superposition of bipy (dark grey), en (orange), phen (grey), tmen (black) and ampy (light grey).

Conclusion

Our results reveal that the reactivity of these complexes increases with the decreasing π-acceptor ability of the N^N chelates: phen < bipy ≈ tmen < ampy < en. Analysis of the superposition of the complexes indicates that the low reactivity of the tmen complex compared to its en analog is caused by the steric hindrance provided by CH_3 groups.

References

[1] M. Chrzanowska, et al. Dalton Trans. 46 (2017) 10264.
[2] G. M. Sheldrick, Acta Cryst. C71 (2015) 3.
[3] M. Chrzanowska, et al. J. Coord. Chem. 71 (2018) 1761.

Figure 1

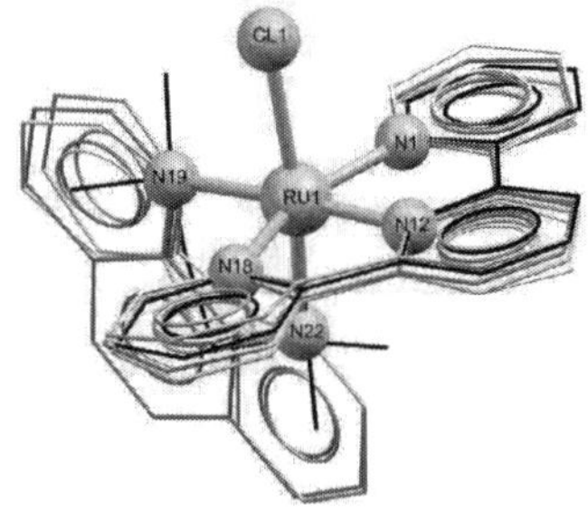

P84

"POSS-based open cage" and non-typical structure of its hydrolytic condensation product

P. Wytrych[1], A. Władyczyn[1], T. Lis[1], Ů John[1]
[1]Wrocław University of Science and Technology, Faculty of Chemistry, Wrocław, Poland

The crystal structures of two inorganic-organic hybrid POSS compounds (POSS = polyhedral oligomeric silsesquioxanes): $Ph_7POSS-(OH)_3$ (**1**) and product of its rearrangement - $Ph_{12}POSS$ (**2**), have been studied using single-crystal X-ray diffraction. Both structures possess peripheral phenyl (Ph) groups directly attached to every silicon atom. Obtained compounds were established according to hydrolytic condensation reactions. The crystal structure of **1** has two components: major one composed of $Ph_7POSS-(OH)_3$ (0.85 s.o.f) (**1a**) and minor one of the formula $Ph_6POSS-(OH)_4$ (0.15 s.o.f) (**1b**). The uniqueness of **1b** derives from the fact, that instead of one phenyl group, there is an additional hydroxyl OH substituent.

In turn, the molecular structure of **2** is the result of a previously undescribed in the literature reaction with **1**, which was catalyzed by DBTL (dibutyltin dilaurate). A fully condensed **2** cage was obtained with high 80% yield. The obtained results may be useful in understanding the further reactivity on the silicon atom among species containing phenyl substituents.

Figure 1

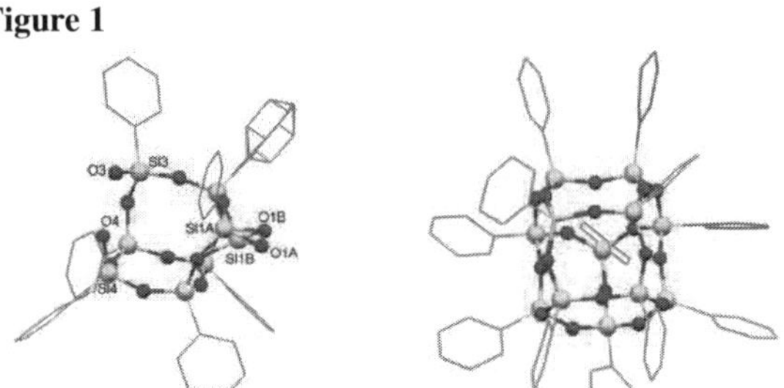

P85

Novel dichloride platinum(II) complexes with 1,2,4-triazolo[1,5-*a*]pyrimidine derivatives and sulfoxide ligands.

M. Jakubowski[1], I. Łakomska[1], K. Kałduńska[2], A. Jarzęcki[3], A. Wojtczak[2]
[1]Nicolaus Copernicus University, Faculty of Chemistry, Toruń, Poland
[2]Nicolaus Copernicus University, Faculty of Chemistry, Department of Biomedical and Polymer Chemistry, Toruń, Poland
[3]City University of New York – Brooklyn College, Brooklyn, NY, United States

Introduction

Novel Pt complexes with triazolepyrimidines and TMSO or DPSO sulfoxides were investigated. Differences in the molecular architecture are important for the anticancer applicability.

Objectives

The aim was to determine the structure of PtCl2 complexes with di-substituted triazolopyrimidines and sulfoxides (**Fig.1.**), and analyse the role of steric effects to the stability of isomers.

Materials and methods

Structures were solved and refined with SHELX-2017 package. [2] DFT calculations of the Torsional Potential Energy Scans along with DPSO-Platinum bonding were performed.

Results

Complexes with TMSO (**1,2**) occur as *cis* isomers, (**3**) is the *trans* isomer.

Fig.1. *Cis*-PtCl2(dmtp)(TMSO) (**1**), *cis*-PtCl2(dbtp)(TMSO) (**2**), *trans*- PtCl2(dbtp)(DPSO) (**3**)

TP ligands are coordinated via N3, similar to complexes reported previously. [2,3] In *trans* isomer (**3**), both Pt-Cl bonds are shorter than these in (**1-2**). In (**3**) the Pt1-S1 distance of 2.2150(18) Å is significantly longer than those of 2.199 Å in (**1-2**). That reflects the steric effect of DPSO larger than that of TMSO. The DFT analysis of the torsion PES for DPSO complexes reveal that the *trans* isomers have shallow minimum corresponding to the solid state conformation. Significantly larger steric effects are found for hypothetical *cis* isomers.

Conclusion

Crystal structure determination and DFT calculations revealed the significance of steric effects of the sulfoxide for the cis/trans isomer stability. Therefore sulfoxides are promising ligands for the synthesis of active anti-cancer agents.

Research performed under BRAIN (Biomedical&phaRmAceutical InterdiscipliNary group)

References

[1] G. M. Sheldrick, Acta Cryst., 2015, C71, 3-8.
[2] K. Hoffmann, J. Wiśniewska, A. Wojtczak, J. Sitkowski, A. Denslow, J. Wietrzyk, M. Jakubowski, I. Łakomska, J. Inorg. Biochem. 172 (2017) 34-45.
[3] Łakomska, M. Babinska, A. Wojtczak, J. Sitkowski, Inorg, Chim. Acta. 453 (2016) 516-521.

Figure 1

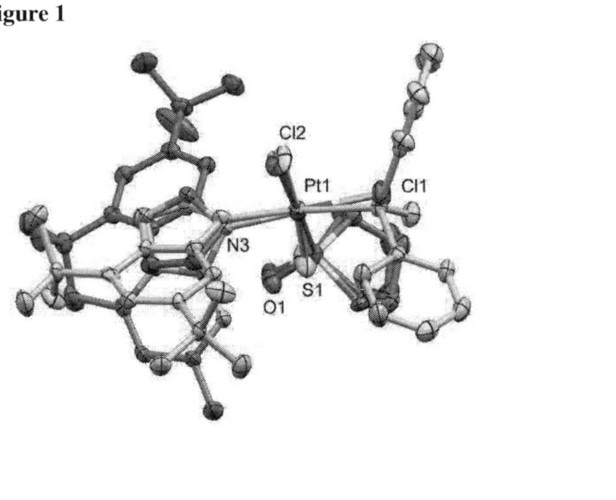

P86

Intermolecular interactions in hypodiphosphate salts of caffeine anf theophylline

M. Otręba[1], K. Ślepokura[1]
[1]Wrocław University of Science and Technology, Faculty of Chemistry, Wrocław, Poland

Hypodiphosphoric acid ($H_4P_2O_6$) is a structural analogue of pyrophosphoric acid ($H_4P_2O_7$), where P^V–O–P^V linkage is replaced by P^{IV}–P^{IV} bond. Pyrophosphates play significant role in cell biochemistry, thus there is a question if hypodiphosphates can replace them. As a continuation of research over hypodiphosphate purinium salts,[1] we decided to investigate interactions in the crystals of caffeinium (1) and theophyllinium (2) hypodiphosphates (Fig. 1). These xanthine derivatives are competitive inhibitors for adenosine receptors.[2]

Figure 1. Scheme of caffeine (1, Caff) and theophylline (2, Teof).

Three new salts have been synthesized and characterized in the crystalline form: (CaffH)($H_3P_2O_6$), $(TeofH)_2(H_2P_2O_6)$ and (TeofH)($H_3P_2O_6$)·H_2O. All crystals were obtained by mixing hypodiphosphate acid with purine. Purity, structures and thermal properties of the compounds were examined by PXRD, XRD and DSC/TGA methods.

All crystals possess hydrogen bond between protonated nitrogen N9 atom and oxygen from hypodiphosphate anion (PP). Hydrogen bonds between neighbouring anions (PP···PP) seem to play important role for caffeinium and hydrated theophyllinium salts. In (CaffH)($H_3P_2O_6$), layers of PP anions are separated by double ribbons of purinium cations (connected only with C–H···O hydrogen bond between $CaffH^+$), while in (TeofH)($H_3P_2O_6$)·H_2O the hypodiphosphate chains are separated by the ribbons of purinium cations and mediating water molecules. In contrast, crystal packing of $(TeofH)_2(H_2P_2O_6)$ is dominated by π–π stacking interactions between $TeofH^+$, with isolated hypodiphosphate anions lying between stacked purinium dimers, similar as previously reported $AdeH^+$.[1] Important role is played by the lone pair–π contacts between hypodiphosphate O atoms and purinium cations.

References

[1] M. Otręba, D. Budzikur, Ł. Górecki & K. A. Ślepokura, *Acta Crystallogr. C.*, **74** (2018) 571.

[2] S. G. Holtzman, S. Mante & K. P. Minneman, *J. Pharmacol. Exp. Ther.*, **256** (1990) 62.

Figure 1

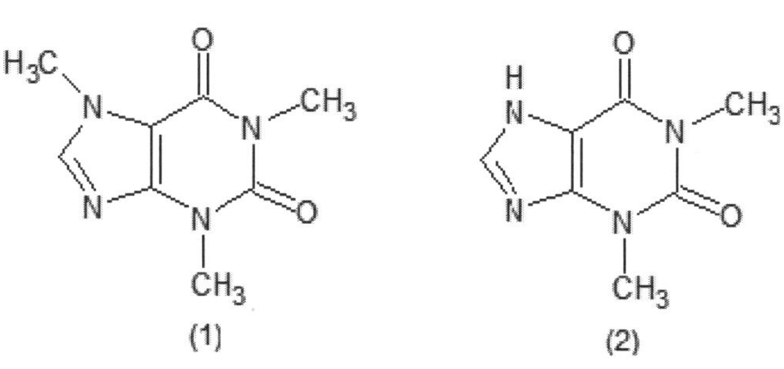

(1) (2)

P87

Analysis of the nature of chemical bonds in fenamic acids salts

M. Krawczyk[1], I. Majerz[1]
[1]Wrocław Medical University, Faculty of Pharmacy, Wrocław, Poland

In order to create drugs that are more effective and exhibit appropriate physical and pharmacological properties, syntheses of salts and coordination compounds were undertaken. New forms of pharmaceuticals of higher solubility, stability, bioavailability and low toxicity are sought. Our interest focuses on obtaining crystals of salts and coordination compounds of fenamic acids, belonging to non-steroidal anti-inflammatory drugs (NSAIDs). In reactions of fenamic acids with selected alkali metal hydroxides and transition metal salts, new drug compounds were obtained. The structures of sodium and potassium salts of mefenamic acid were studied previously [1], but other polymorphic forms were also obtained. There are known structures of meclofenamic acid coordination compounds with copper(II) and cadmium(II) [2]. Our previous studies of the structure of sodium diaquafenamate–water revealed that the strength of Na-O bonds is similar to the weak intramolecular hydrogen bond [3]. The method of investigation of the nature of chemical bonds is a theoretical analysis of electron density (QTAIM) [4] and noncovalent interaction (NCI) [5] performed for the experimental crystal structures. We have obtained series of salts of fenamic acids with different cations to improve physical and pharmaceutical properties and to characterize the weak interactions and chemical bonds linking the cations with fenamic acids.

[1] R. Kruszynski, A. Trzesowska-Kruszynska, P. Majewski, E. Łukaszewicz, K. Majewska, T. Sierański, B. Lewiński *J.Mol. Struct.* 2010, *970*, 79–89.
[2] D. Kovala-Demertzi, M. Staninska, I. Garcia-Santos, A. Castineiras, M. A. Demertzis *J. Inorg. Biochem.* 2011, *105*, 1187–1195. [3] M.S. Krawczyk, I. Majerz, *Acta Cryst.* 2019, *B***75**, 766-774
[4] R.F. W. Bader, (1990). *Atoms in Molecules: A quantum Theory*, Oxford University Press, New York.
[5] J. Contreras García, E.R. Johnson, S. Keinan, R. Chaudret, J.-P. Piquemal, D.N. Beratan, W. Yang, *J. Chem. Theory Comput.* 2011, *7*, 625–632.

P88

The structures of 3,5-diiodo-l-tyrosinato copper(II) complexes stabilized by weak interaction in and around the metal centers

A. Wojciechowska[1], T. Rojek[1], A. Gągor[2]
[1]Wrocław University of Science and Technology, Faculty of Chemistry, Wrocław, Poland
[2]Institute of Low Temperature and Structure Research, Polish Academy of Sciences, Wrocław, Poland

The preparation of metal complexes in the crystalline state has become the most important subject of research in the fields related to engineering, chemistry as well as physics due to the potential application of coordination compounds as magnetic materials, radiation detectors, piezoelectric or laser hosts [1].

A hybrid 3,5-diiodo-l-tyrosinato copper(II) complexes of formulae [Cu(l-I2TyrOH)Cl(phen)]·2H2O (**1**) and [Cu(l-I2TyrOH)(phen)(H2O)]·NO3 (**2**) (l-I2Tyr = 3,5-diiodo-l-tyrosine, phen = 1,10-phenanthroline) crystalize in trigonal system (*R*3:*H* space group) with the unit cell dimensions $a = b = 27.1560(5)$ Å, $c = 8.5857(3)$ Å (Z=9) and monoclinic *P*21 space group with cell dimension $a = 8.1218(3)$ Å, $b = 9.3200(3)$ Å, $c = 15.3403(5)$ Å (Z=2), respectively. The geometries of resulting {CuN3OCl} and

{CuN3O2} coordination spheres can be described as distorted square pyramid with all N atoms and carboxylate O atom located in the corners of square base and chloride ion or water molecule in the apex of **1** and **2**, respectively. The values of Adison τ parameter of 0.357 (**1**) and 0.335 (**2**) indicate the relatively small distortion of the square pyramids. The l-I2TyrO- entity is bent and therefore located at the opposite site to the chlorine ion (**1**) or water molecule (**2**). As a result the coordination units adopt characteristic scorpion-shaped architecture with L-I2TyrO- side chain aromatic ring in its tail. The scorpion-shaped arrangement is supported by formation of stabilizing $\pi\cdots\pi$ stacking and metal$\cdots\pi$ interaction between the tail and aromatic moiety of phen ligand and Cu(II) center, respectively.

[1] a) X.-S. Gao, H.-J. Dai, Y. Tang, M.-J. Ding, W.-B. Pei, X.-M. Ren, *ACS Omega* 4 (2019) 12230-12237. b) P. Thomas, R. Junjuri, N. Joy, M. Siemer, M. K. Gundawi, R. Philip, K. Al-Shamery, G. P. Joseph, *J. Mater. Science : Materials in Electronics* 30 (2019) 1-15.

Figure 1

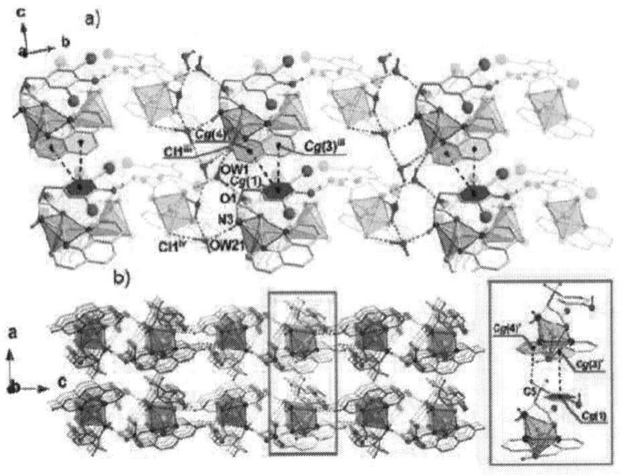

Figure 1. The role of $\pi\cdots\pi$ interactions in arrangement of coordination units into three-dimensional supramolecular network in the structure of a) 1 and b) 2. All C-bounded H-atoms and iodine atoms (picture b) are omitted for clarity.

P89

Copper(II) complexes with L-arginine

A. Wojciechowska[1], T. Rojek[1], A. Gągor[2]
[1]Wrocław University of Science and Technology, Faculty of Chemistry, Wrocław, Poland
[2]Institute of Low Temperature and Structure Research, Polish Academy of Sciences, Wrocław, Poland

From the success of *cis*-platin in the treatment of cancer increasing attention has been paid to metal-based compounds. The copper(II) complexes with l-arginine (l-Arg) present strikingly good biological properties [1]. We prepared single crystals of new complexes of formulae [Cu(*m*-O,O"-NO3)(l-Arg)(bpy)]NO3 (**1**) and [CuCl(l-Arg)(bpy)]Cl·3H2O (**2**) (bpy=2,2"-bipyridine).

The compound **1** crystalizes in orthorhombic *P*212121 space group with cell parameters: *a*=6.8148(6) Å, *b*=13.8419(9) Å, c=23.3427(17) Å. The coordination sphere of copper(II) center shows distorted octahedral geometry with shorter Cu-Ol-Arg, N distances in the basal plane. The values of apical Cu-ONO3 bond lengths indicate elongation of octahedron (*T*=0.751). The complex (**1**) crystalized as a one-dimensional coordination polymer by means of NO3- ions. Such an arrangement allows the guanidine N atoms from neighbor chains to act as donors in the N–H···O(NO3) hydrogen bonds resulting in the formation of a layer. The N–H···O interaction between guanidine group and carboxylate O atom links adjacent layers.

The compound **2** crystalizes in triclinic noncentrosymetric *P*1 space group with cell parameters : *a*=8.3645(9) Å, *b*=10.2116(10) Å, c=14.9467(16) Å, a=84.588(9), β=81.461(9), γ=67.108(10). The pentacoordinate environment around Cu2+ ion is provided by two bpy N atoms, Cl- anion and the amino N and carboxylate O atoms from l-arg giving {CuN3OCl} core. The coordination sphere can be described as distorted square pyramid (*t*=0.11). The [CuCl(l-Arg)(bpy)Cl]+ cations are linked in dimeric units which are further interconnected to form 1D *zig-zag* chains. The lattice Cl- anions and water molecules fill the spaces between chains and connect them into 2D layers. The amino and guanidine groups from l-Arg moiety hydrogen bonded to lattice water molecule contribute to the stabilization of the 3D architecture.

[1] A. K. Patra, T. Bhowmick, S. Roy, S. Ramakumar,A. R. Chakravarty, *Inorg. Chem.* **2009**, 48, 2932.

Figure 1

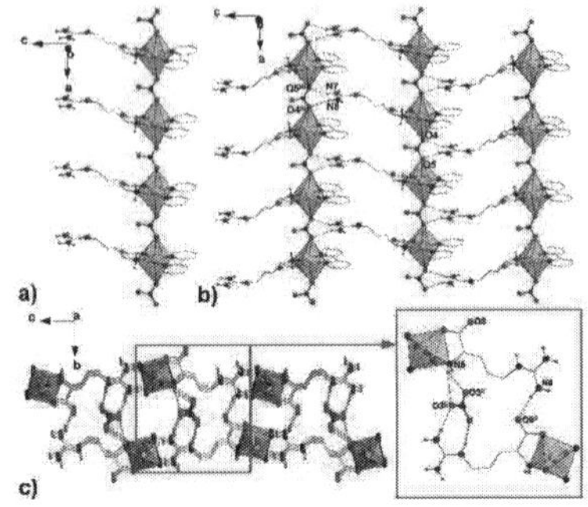

Figure 1. Crystal packing for the coordination polymer 1. a) One-dimensional polymeric chain running along the *a* axis (b) 2D layer parallel to the (101) plane; (c) packing of 2D layers shown in yellow and light grey). All H atoms not involved in the creation of hydrogen bonds are omitted for clarity.

Figure 2

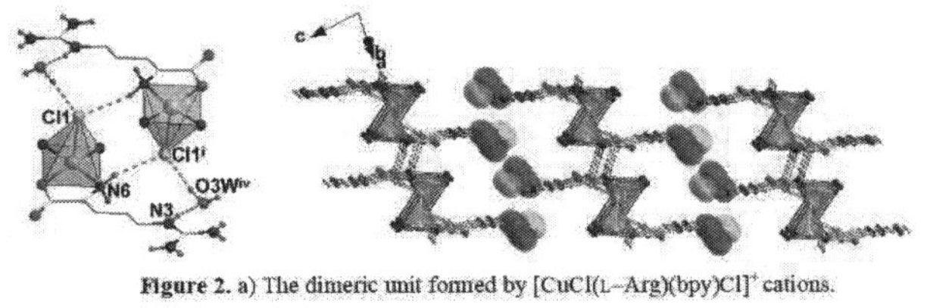

Figure 2. a) The dimeric unit formed by [CuCl(L-Arg)(bpy)Cl]+ cations. b) The view of 3D supramolecular network. All H atoms not involved in the creation of hydrogen bonds are omitted for clarity.

P90

3D coordination polymer of silver(I) bridged by 1,4-dicyanobenzene and nitrate(V) ions.

K. Gutmańska[1], A. Ciborska[1], A. Dołęga[1]
[1]Gdansk University of Technology, Department of Inorganic Chemistry, Gdańsk, Poland

Introduction

Though nitriles are weakly basic and the nitrile ligands are often labile, a certain number of 1,4-dicyanobenzene (1,4-DCB) and other nitrile complexes of metal ions were isolated as solids e.g. [1,2].

Objectives

We are currently studying the polynuclear complexes of silver(I) with the metal ions bridged by aromatic ligands such as 1,2-di(4-pyridyl)ethylene. Our major goal is isolation of stable species of various nuclearity to learn how their physicochemical properties depend on the length of the chain. Lately we have decided to use 1,4-DCB for the synthesis and the initial results are presented in this poster communication.

Materials and methods

The compound **I** crystallized from EtOH/toluene. Colourless plates were suitable for the X-ray measurement carried on STOE IPDS 2T at 120 K. Data collection and image processing was performed with X-Area 1.75 [3]. Structure of **I** was solved using the SHELX–2014 [4].

Results

With a very simple system i.e. silver(I) nitrate and 1,4-DCB we obtained the 3D CP presented in the figure below. The compound crystallizes in *P*21/*c* group of the monoclinic system. The structure was solved to the discrepancy factor $R_1 = 0.0181$ ($R_{int} = 0.0244$). The asymmetric unit contains one silver atom, one nitrate(V) anion and a half of the 1,4-DCB molecule. The silver ions feature C.N.=5

Conclusion

Simple system consisting of silver nitrate, 1,4-dicyanobenzene in an appropriate mixture of solvents returns 3D CP. We intend to increase the size of the anion in a step-wise manner to reduce the dimensionality of the polymer and compare the properties of the CPs.

Fig. 1 Molecular structure of compound I with thermal ellipsoids at 50%.

[1] C. J. Holler, K. Muller-Buschbaum, Inorg. Chem., 47 (2008) 10141.

[2] B. Nohra, R. Reau, C. Lescop, Eur. J. Inorg. Chem., (2014) 1788.

[3] STOE & Cie GmbH, X-Area 1.75, (2015) Darmstadt.

[4] G. M. Sheldrick, (2014) SHELXL-2014; University of Göttingen and Bruker AXS: Karlsruhe.

Figure 1

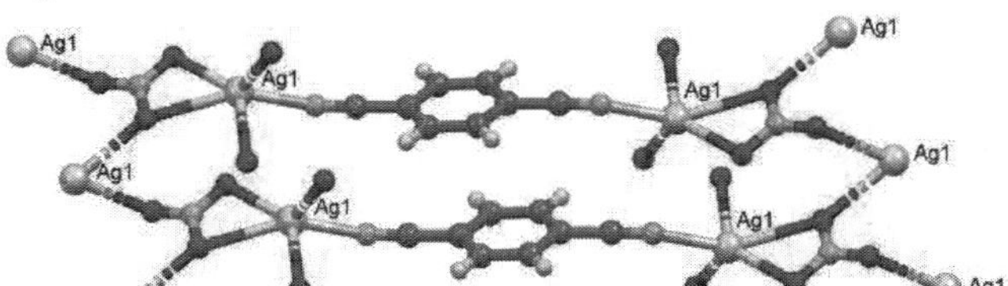

P91

Crystal structures of 4-ethynyl-2,1,3-benzothiadiazole and 4,7-diethynyl-2,1,3-benzothiadiazole

J. Alfuth[1], K. Kazimierczuk[1], T. Olszewska[1]

[1]Gdansk University of Technology, Gdańsk, Poland

Molecules containing 2,1,3-benzothiadiazole unit have been systematically studied due to their unique physical and chemical characteristics. They have found application in organic light-emitting diodes, solar cells, liquid crystals and many others[1] and are a useful model in crystal engineering for weak intermolecular interaction investigations. In this contest, we have focused our attention on compounds **1** and **2** bearing the acetylene unit at position 4- or 4- and 7-, respectively (Figure 1). Both positions are of high importance for photo-optical properties of 2,1,3-benzothiadiazoles.[2] The objective was to investigate the influence of acetylene moiety on crystal architectures of compounds **1** and **2**.

Crystals of both derivatives were prepared by dissolving them in ethyl acetate and then letting the solvent to slowly evaporate. X-Ray diffraction analysis was performed with a STOE IPDS 2T diffractometer with graphite-monochromated Mo-*Kα* radiation (λ = 0,71073 Å).

Molecules of **1** in its crystal structure are connected *via* ≡C–H···N hydrogen bonds (d = 2.44 Å, θ = 178°) forming infinite chains (Figure 2A). The stacking of chains occurs in antiparallel fashion with π-π stacking interactions between the rings. In contrast, molecules of **2**, bearing two ethylene moieties, interact *via* strong chalcogen bonds in the form of cyclic four-membered supramolecular synthons $[S\cdots N]_2$ (Figure 2B), which is rarely seen in this class of compounds.[3] Hydrogen bonds are present as well but as weak ≡C–H···π(ethynyl) interactions between two adjacent molecules.

In conclusion, our study shows that the main driving forces responsible for crystal architecture are switched from weak hydrogen intermolecular interactions present in **1** to strong chalkogen bonds in **2** by simple introduction of an additional acetylene substituent.

[1] Neto B.A.D. et al. *J. Org. Chem.* 2013, 228–255.

[2] da Cruz E.H.G. et al. *New J. Chem.,* 2014, 2569–2580.

[3] Cozzolino A.F. et al. *Chem. Soc.* 2005, 127, 3184–3190.

Figure 1

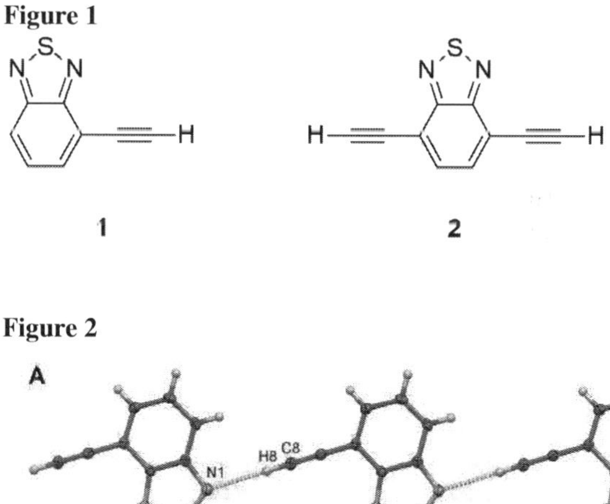

1 **2**

Figure 2

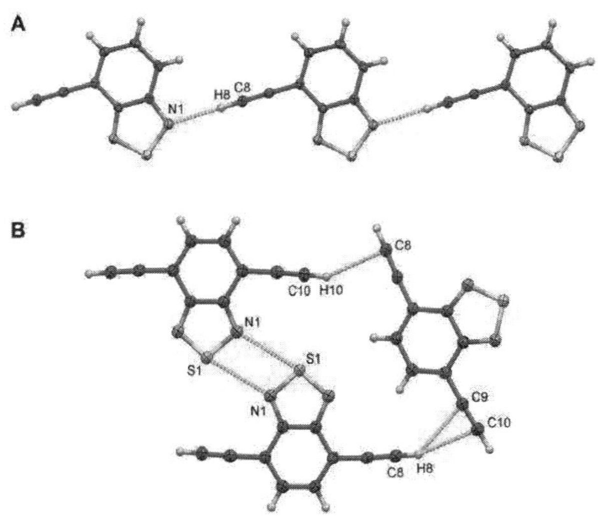

P92

Purine alkaloid systems with 2,6-dihydroxybenzoic acid as a coformer – X-ray structural analysis and solubility studies.

M. Gołdyn[1], M. Pawlaczyk[1], E. Bartoszak-Adamska[1]
[1]Adam Mickiewicz University, Department of Chemistry, Poznań, Poland

One of method to improve the solubility of active pharmaceutical ingredients (APIs) is a cocrystal or salt formation with well-selected, pharmaceutically acceptable coformer[1,2]. The drug solubility improvement of poorly soluble API is very important from a pharmaceutical industry perspective, because this parameter may affect the bioavailability and dosage of the drug[3].

2,6-dihydroxybenzoic acid (γ-resorcylic acid) was chosen as a coformer to cocrystallization with purine alkaloids, like theobromine, theophylline and caffeine. The goal was to synthesize various alkaloid-γ-resorcylic acid systems by crystallization from solution and by neat or liquid-assisted grinding. A single-crystal (SXRD) and powder X-ray diffraction (PXRD) measurements were performed. UV-Vis spectroscopy was used to determine solubility of obtained purine alkaloid derivatives.

Theobromine with γ-resorcylic acid form salt hydrate or salt depending on the crystallization conditions. Cocrystallization of theophylline and this acid only leads to the salt hydrate formation. Caffeine with this acid form salt (Fig. 1). The SXRD method confirmed proton transfer between given alkaloid and γ-resorcylic acid. The PXRD studies have shown the possibility of these compounds synthesis by grinding method. The solubility of alkaloids after cocrystallization was improved.

Fig. 1. Results of theobromine, theophylline and caffeine cocrystallization with 2,6-dihydroxybenzoic acid.

References

[1] B. Sarma, B. Saikia, *CrystEngComm*, 2014, 16, 4753-4765.
[2] P. Vishweshwar, J. A. McMahon, J. A. Bis, M. J. Zaworotko, *J. Pharm. Sci.* 2006, 95, 499-516.
[3] P. Sanphui, A. Nangia, *Journal of Chemical Sciences*, 2014, 126, 1249-1264.

Acknowledgement
The work was supported by grant no. POWR.03.02.00-00-I026/16 co-financed by the European Union through the European Social Fund under the Operational Program Knowledge Education Development

Figure 1

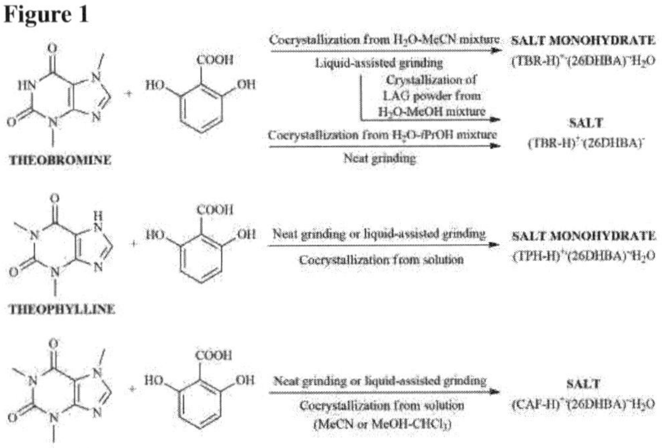

P93

Natural Bond Orbital Approach to the Intermolecular Interactions

R. Kruszyński[1]
[1]Łódź University of Technology, Department of X-ray Crystallography and Crystal Chemistry, Łódź, Poland

The different approaches to explain and categorise the different intermolecular interactions were used throughout the time [1]. Most of them were based on various simplifications, allowing comparison of diverse interactions character in easy-to-understand terms. The historically one of oldest, but less commonly used method is based on Natural Bond Orbitals (NBOs) [2]. The NBO theory is based on assignment of electron density into idealised Lewis structures (non-bonding electron pair and covalent 2 electron bond) and thus it allows simple interpretation of complex interactions. Pure Lewis approach denies existence of any non-covalent interaction, but the NBO method takes into account "delocalisation" of electrons, i.e. the transfer of some electron density from Lewis orbital into other empty orbitals, in terms of second order perturbation theory. Consequently, the four constructs are considered: the 1-center orbital (equivalent to non-bonding electron pair), 2-center orbital (equivalent to covalent 2 electron bond), anti-bonding orbitals and Rydberg-type orbitals. Such approach, however still limited to presence of only 2 centre molecular orbitals (real 3 or more centre orbitals are treated as interactions of above mentioned constructs), is useful in categorisation of intermolecular interactions as interactions between the specific orbitals, between which the explicit electron interactions occurs. The different intermolecular interactions (hydrogen bonds, halogen and stacking interactions, etc.) are presented, compared and discussed in terms of interactions between natural orbitals.

References

[1] Desiraju, G. R. & Steiner, T. (1999). The Weak Hydrogen Bond in Structural Chemistry and Biology. IUCr Monograph on Crystallography 9. Oxford University Press.
[2] Reed, A. E., Curtis, L. A. & Weinhold, F. A. (1988). Chem. Rev. 88, 899-926

P94

Similarities in diversity - crystal structures of glycyrrhetinic acid acetate derivatives

B. Wicher[1], D. Langer[1], E. Tykarska[1]
[1]Poznan University of Medical Sciences, Department of Chemical Technology of Drugs, Faculty of Pharmacy, Poznań, Poland

Introduction

Glycrrhetinic acid (GE) belongs to pentacyclic triterpenoids, and exhibits, as well as its derivatives many biological activities [1]. Our previous studies of GE esters showed that despite significant differences in alkyl substituents their crystal structures were similar [2].

Objectives

The goal of our research was the synthesis of GE derivatives with no strong hydrogen bond donors to determine whether the similarity of crystal packing will be maintained.

Materials and methods

Derivatives **2-5** (Fig. 1) were obtained through alkylation of glycyrrhetinic acid acetate (**1**) with a proper alkyl halide and subjected to crystallization and single crystals X-ray analysis.

Results

Depending on the alkyl substituent, a variety of molecular packing modes in crystal structures are observed. However, despite these differences, all structures are built of layers, the construction of which is based on ribbons. In all 1-D motifs, triterpene molecules are arranged in a head-to-head fashion.

In crystals **2** and **4**, the ribbons are nearly perpendicular to the base of the layers (Fig. 2a), but in the latter, the adjacent 1-D motifs are in a head-to-head relation, whereas in **2** in a head-to-tail manner. A different construction of the layer is observed in **3**. Two symmetry independent ribbons arranged in a head-head-tail-tail sequence form a layer (Fig. 2c). In **2**, **3**, and **4** alkyl substituents are directed outside the layer. In **5**, the largest ethylmorpholine substituents become an integral part of the layer as they are placed between two adjacent head-to-tail arranged ribbons (Fig. 2d).

Conclusion

Terpenoid derivatives that do not have groups capable of forming strong hydrogen bonds form a variety of crystal structures. Even so, it is possible to distinguish recurring structural motifs, which are layers composed of ribbons.

References

[1] Liu J. *J., Ethnopharmacol.* 2005, **100**, 92.
[2] Langer D., et. al. *Acta Cryst.* 2016, **B72**, 584.

Figure 1

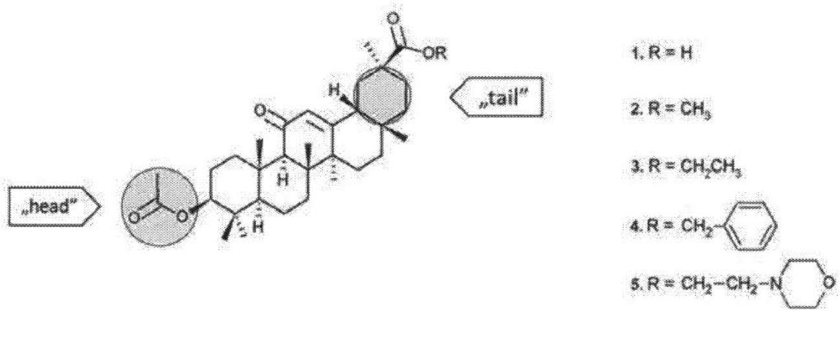

Figure 2

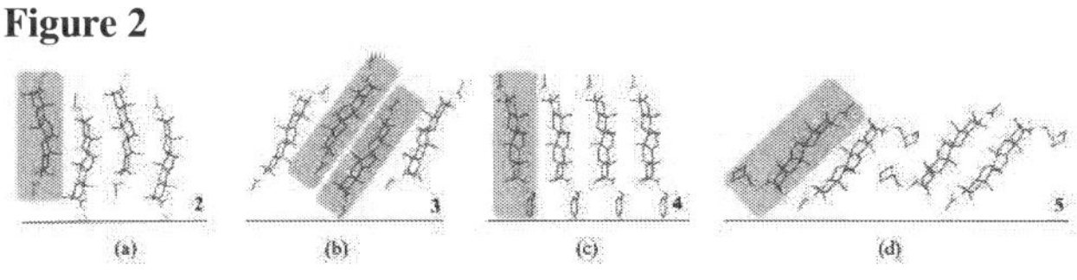

P95

Crystallographic and NMR structuralstudies on halogen derivatives of 2-deoxyglucose and 2-deoxymannose

M. Ziemniak[1], M. Sołtyka[2], D. Trzybinski[1], S. Pawlędzio[1], A. Zawadzka-Kazimierczuk[1], W. Kozminski[1], W. Priebe[3], K. Wozniak[1], B. Pajak[2]
[1]University of Warsaw, Department of Chemistry, Warsaw, Poland
[2]The General Karol Kaczkowski Military Institute of Hygiene and Epidemiology, Independent Laboratory of Genetics and Molecular Biology, Warsaw, Poland
[3]MD Anderson Cancer Center, The University of Texas, Department of Experimental , Houston, United States

Introduction

In this on-going work we present crystallographic and NMR studies on several halogen derivatives of 2"-deoxyglucose and 2"-deoxymannose. These compounds are competitive inhibitors of hexokinase, one of the key regulators of glycolysis, which is considered as a potential target in anticancer therapy. It is expected that substitution of 2"H atom with a halogen may improve the inhibitory potential of these compounds.

Objectives

To crystallise a small set of derivatives of 2"-deoxyglucose and 2"-deoxymannone to compare their structural features in solid state and in the solution, using NMR spectroscopy as a complementary method.

Materials and methods

Diffraction data were collected on the Agilent Technologies SuperNova Dual Source diffractometer with the CuKα radiation. The structural determination was carried out using the SHELX package. Molecular interactions in the crystals was identified using PLATON program.

Results

Five derivatives of 2"-deoxyglucose and 2"-deoxymannone were crystallised and their crystal structure were determined. crystallographic analysis showed that replacement of 2"hydrogen by a halogen atom of various size (F, Cl and I) caused changes in the internal geometry of molecules and could lead to appearance of specific hydrogen bonds. The NMR studies using correlation experiments (NOE, COSY) reveal that halogen substitution alter their conformational equilibria in the solution. These initial data have been compared and on-going theoretical calculation are expected to provide better insight into the influence of the halogen atom on the structure and energetics of these compounds.

Conclusion

We expect that better understanding of the structural features of these halogen derivatives of 2"-deoxyglucose and 2"-deoxymannone may aid, along with biological studies, in explaining the mechanism of binding, and possibly in developing of better inhibitors. **Financing**: Projects financed by NCN: UMO-2017/25/B/NZ3 /00251UMO-2017/24/C/NZ1/00366

P96

Synthesis and structural studies of 5-methylamine derivatives of 4-aminoaryl-2-phenyl-6-methylpyrimidine

I. Bryndal[1]
[1]Wrocław Medical University, Faculty of Pharmacy, Department of Drugs Technology, Wrocław, Poland

Pyrimidine core is found in nucleotides, thiamine, uric acid, alloxan, barbiturates and HIV drugs. Since its discovery and laboratory synthesis, the pyrimidine derivatives have become the essential structural unit of wide range pharmaceuticals having important biological activities [1].

The present paper is a continuation of our earlier works about the synthesis, biological activity and structural studies of 6-methyl-2-phenyl-5-substituted pyrimidine derivatives. Microbiological testing, designed and synthesized as immunomodulating agents, showed them to possess antibacterial and antifungal activities [2]. Additionally, the presence of a hydroxy group in 5-position of the pyrimidine ring significantly increases its cytotoxicity towards both, i.e. the cancer (HeLa, K562 and CFPAC) and normal (HUVEC) cell lines [3].

As part of our ongoing studies of this class compounds, we have synthesized 5-methylamine derivatives from ethyl 4-methyl-2-phenyl-6-thioxo-1,6-dihydropyrimidine-5-carboxylate [6] as a starting material, namely, N-(4-chlorophenyl)-5-[(ethylamino)methyl]-6-methyl-2-phenylpyrimidin-4-amine (5a), N-(4-ethoxyphenyl)-6-methyl-2-phenyl-5-[(prop-2-en-1-ylamino)methyl]pyrimidin-4-amine (5b) (Fig. 1) and N-(4-

methyphenyl)-5-[(4-methylpiperazin-1-yl)methyl]-2-phenylpyrimidin-4-amine (5c). The compounds were characterized by 1H-NMR, IR and mass spectroscopies. The molecular structures were further studied by single-crystal X-ray diffraction and obtained structural data were compared with the aromatic amine derivatives described earlier.

[1] R. Dua, S. Shrivastava, S.K. Sonwane, S.K. Srivastava, Adv. Biol. Res. 5 (2011) 120.
[2] J. Cieplik, M. Stolarczyk, J. Pluta, O. Gubrynowicz, I. Bryndal, T. Lis, M. Mikulewicz, Acta Poloniae Pharmaceutica, 72 (2015) 53.
[3] M. Stolarczyk, I. Bryndal, A. Matera-Witkiewicz, T. Lis, K. Królewska-Golińska, M. Cieślak, J. Kaźmierczak-Barańska, J. Cieplik, Acta Cryst. C74 (2018) 1138.

Figure 1

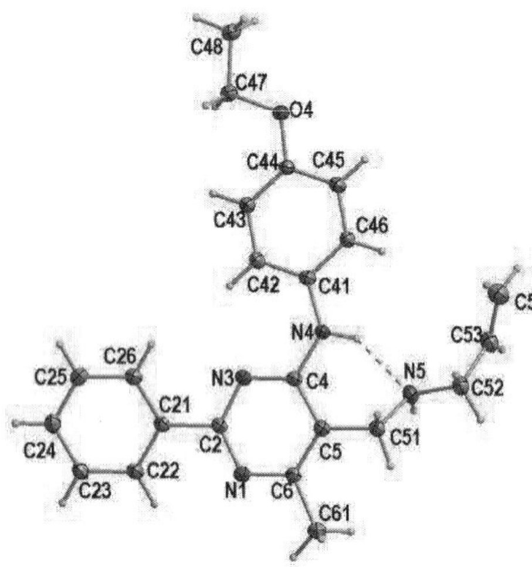

P97

Analysis of influence of anions and solvent molecules on the structure of imidazolium salt cations

A. K. Gzella[1], J. Garbarczyk[2], S. Chmielewska[3]
[1]Poznan University of Medical Sciences, Department of Organic Chemistry, Poznań, Poland
[2]Hipolit Cegielski University of Applied Sciences Gniezno, Gniezno, Poland
[3]Poznan University of Technology, Poznań, Poland

The aim of the study is to analyse the relationship between the cation conformation and the kind and size of the anion, and - in some cases - the additional presence of the solvent molecules in the selected imidazolium salt crystals as a biologically active compounds.

The studies were conducted for salts composed of 1-benzyl-3-cyclododecyloxy-methylimidazolium and 1-benzyl-3-cyclododecyloxymethyl-2-methylimidazolium cations and Cl^-, Br^-, $NO3^-$, $BF4^-$, $PF6^-$, $CH3COO^-$, $CF3COO^-$ anions, based on WAXS (Wide Angle X-ray Scattering), X-ray Diffraction methods, differential scanning calorimetry (DSC) method and quantum chemistry (QC) calculations and molecular mechanics modeling (MM).

The analysis of the influence of anion and water molecules on the conformation of 1-benzyl-3-cyclododecyloxymethylimidazolium (I) and 1-benzyl-3-cyclododecyloxymethyl-2-methylimidazolium (II) cation molecules based on the X-ray method showed that anions react stronger with a 1-benzylimidazolium or 1-benzyl-2-methylimidazolium fragment than with the cyclododecyloxymethyl, and that the ion-ion electrostatic interaction for the salt with the 1-benzyl-2-methylimidazolium cation moiety is weaker than for the salt with the 1-benzylimidazolium cation fragment. MM calculations conducted for 1-benzyl-3-cyclododecyloxymethylimidazolium and 1-benzyl-3-cyclododecyloxymethyl-2-methylimidazolium cations showed that their conformational change is easiest with the torsion angle t1 associated with the rotation of the benzyl system, and conformational change energy barrier for a cation with an additional methyl substituent at the C-2 position of the imidazolium system is lower than for a cation without this substituent.

It is believed that the information obtained on the supramolecular architecture of the tested salts can be used in modern organic chemistry, biology and medicine.

Figure 1 *Cation molecules I (R=H) and II (R=CH3); t1, t2, t3 – torsion angles associated with the rotation of relevant parts of the molecule.*

Figure 1

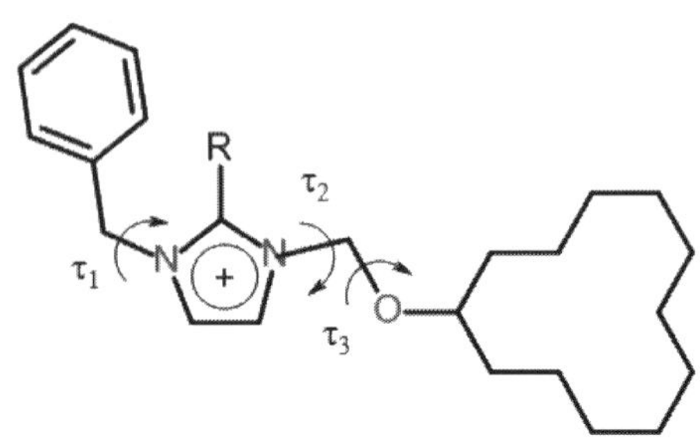

P98

AMINE-IMINE TAUTOMERISM IN ISOMERIC 2-(*o*-,*m*-,*p*-HYDROXYPHENYL)-AMINO-1,3-THIAZOL-4(5*H*)-ONES

A. Pyrih[1], A. K. Gzella[2], R. Lesyk[3], M. Jaskólski[1,4]
[1]Adam Mickiewicz University, Department of Crystallography, Faculty of Chemistry, Poznań, Poland
[2]Poznan University of Medical Sciences, Department of Organic Chemistry, Poznań, Poland
[3]Danylo Halytsky Lviv National Medical University, Department of Pharmaceutical, Organic and Bioorganic Chemistry, Lviv, Ukraine
[4]Polish Academy of Sciences, Center for Biocrystallographic Research, Institute of Bioorganic Chemistry, Poznań, Poland

This work continues our studies of amine-imine tautomerism in the pharmacologically important derivatives of 2-phenylamino-1,3-thiazol-4(5H)-one [1].

The aim of this investigation has been to unambiguously clarify the structure of three isomeric 2-(o-,m-,p-hydroxyphenyl)-amino-1,3-thiazol-4(5H)-ones with respect to their tendency to undergo the aforementioned tautomerism. The experiments were carried out in the solid state using single crystal X-ray diffraction.

The structural results reveal that the molecules of all the investigated compounds adopt the amine form (A-1 structure, Fig. 1) as well as the same synperiplanar conformation.

For all the investigated compounds the amidine H atoms were located on the difference Fourier maps at the exocyclic N6 atom. In the crystals of 1 and 2, the amine N6 atom forms intermolecular H-bonds with the exocyclic O14 atom as the acceptor, N6-H6···O14i. In the case of crystal 3, the molecules are connected via intermolecular H-bonds N6-H6···N3i, in which both amidine nitrogen atoms (N3 and N6) participate (Fig. 2).

Additional evidence corroborating the correctness of the assignment of the tautomeric form is provided by the analysis of the bond lengths C2-N3 and C2-N6. These bonds tend to have comparable length in the case of the amine form, whereas for the imine form they retain their single/double bond character.

Figure 1. *Possible tautomeric structures of compounds **1** – **3**.*

Figure 2. *Molecular structures of **1** - **3***

References

M. Kowiel, Ph.D. Thesis (2015), Poznan University of Medical Sciences

Figure 1

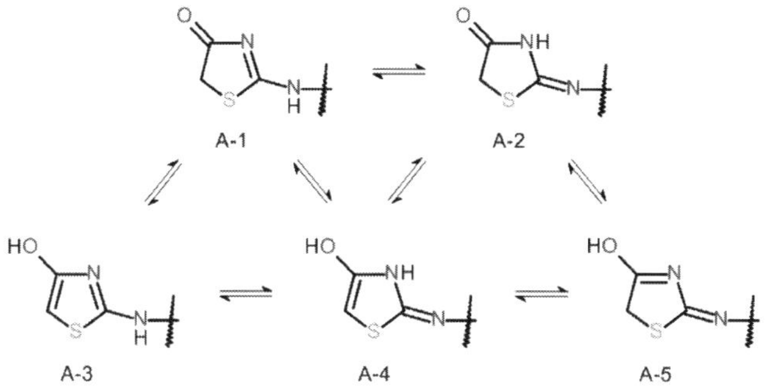

Figure 2

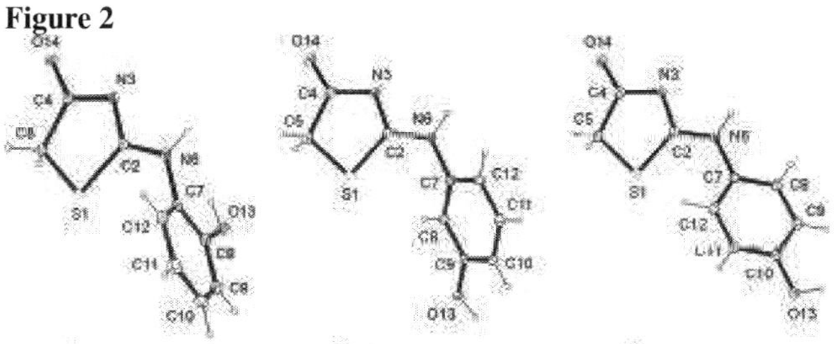

1 2 3

Bio-Crystallography III: Instrumentation & hybrid methods

P99
EMBL Beamlines for Macromolecular Crystallography at PETRA

T. R. Schneider[1], G. Bourenkov[1], G. Pompidor[1], D. von Stetten[1], I. Bento[1], M. Aghte[1], S. Panneerselvam[1]
[1]European Molecular Biology Laboratory (EMBL), Hamburg, Germany

EMBL-Hamburg operates two MX beamlines, P13 and P14, at PETRA III (DESY, Hamburg).

P13 delivers high photon fluxes at energies down to 4 keV with typical data collection times of 5 min. S-SAD phasing is achieved routinely [1] also from multiple crystals [2]. Data collection at low energies does not require any special preparation of the sample, i.e. a crystal attached to a standard SPINE-pin can be automatically mounted onto the goniostat.

P14 can provide a collimated rectangular homogeneous beam that can be shaped to any size between 10 and 200 µm (crystal life-time ~2 minutes at 100 K) or a micro-focus beam (5 µm , crystal life-time ~500 ms at 100 K). The collimated beam can be used to illuminate large (50-200 µm) and small crystals homogeneously and/or to resolve diffraction from large (>1000 Å) unit cells. Prominent recent applications: a set of structures of the human 20S proteasome [3], the crystal structure of the mediator complex [4], structural enzymology at atomic resolution [5], serial data collections on cryogenically cooled crystalline suspensions [6] or *in situ* from CrystalDirectTM crystallization plates.

Since 2018, a second endstation "T-REXX" is in operation on P14 providing an open environment for the implementation of custom pump-probe time-resolved experiments.

The beamlines are embedded in the Integrated Facility for Structural Biology that offers access to up-stream service such as characterization of samples prior to crystallization, high throughput crystallization, and automatic crystal harvesting with a CrystalDirectTM Harvester.

More information including access modalities can be found at: www.embl-hamburg.de/services/mx.

References

[1] Freire et al. (2019) Mol Cell. 73:1282
[2] Cianci et al. (2019) Acta Cryst. D75:192
[3] Schrader et al. (2016) Science 353:594
[4] Nozawa et al. (2017) Nature 545:248 [5] Dai et al. (2019) 573:609 [6] Gati et al. (2014) IUCrJ 1:87.

P100
X-ray crystallography, used to gather spatial knowledge about biomolecules, requires several handling steps. Novel sample holder act as a platform supporting all steps from crystallization, ligand soaking and data collection at both, ambient and cryogenic temperature without direct crystal handling.

C. Feiler[1], D. Wallacher[2], I. Sarrou[3], U. Müller[3,4], M. S. Weiss[1]
[1]Helmholtz-Zentrum Berlin für Materialien und Energie GmbH, Macromolecular Crystallography, Berlin, Germany
[2]Helmholtz-Zentrum Berlin für Materialien und Energie GmbH, Sample Environment, Berlin, Germany
[3]Center for Free Electron Laser Science, DESY, Hamburg, Germany
[4]Max IV Laboratory, BioMax, Lund, Sweden

Structure determination of protein macromolecules requires the growth of high-quality crystalline material.

Various types of sample holders have been developed and are widely used in protein X-ray crystallography.

Collection of diffraction data from a single protein crystal involves a number of manual handling steps, including crystal fishing, cryoprotection and mounting onto an individual sample holder. Numerous enhancements have been done in order to improve many handling steps, including *in-situ screening* to characterize well-diffracting crystals.
We have developed a novel type of sample holder, which acts as an all-in-one platform. As a replacement of commonly used coverslips, it supports all steps in the workflow from crystal setup to data collection without any direct crystal handling. Further, crystal manipulation is realized in-place and does not require any handling steps of individual crystals. Diffraction data collection can be carried out at both, ambient and cryogenic temperature. Additionally, the sample holder is compatible with SPINE vials, therefore, be used with automatic sample mounting robots. Finally, using the new type of sample holder, the number of crystals/dewar can be drastically increased.
With the latest developments, we have shown that this novel sample holder could be successfully used as fixed target support for serial synchrotron crystallography (SSX). Further developments allow fragment screening campaigns using the latest developed type.

The patent-protected sample holder can be used on both, 22- and 18-millimeter standard 24-well plates. Auxiliary tools support a straightforward workflow, using the novel sample holder with a minimized chance of damaging sensitive crystalline material.

Figure 1

Figure 2

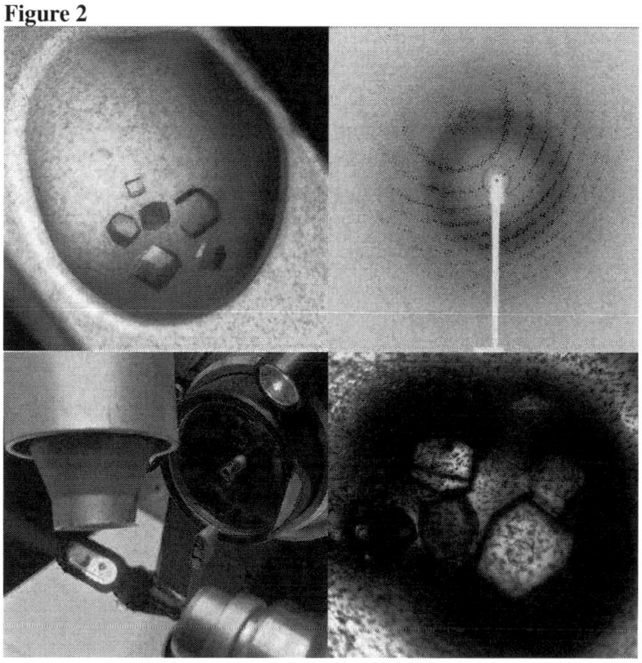

building blocks, nanoparticles will be assembled with high precision into mesoscale materials with optical properties that emerge from interactions between the components. With the recent advances in computational redesign of protein containers, it is now possible to combine these results with nanoparticle synthesis and protein crystallography. By using an innovative design approach with two oppositely charged protein containers as building blocks, a new type of protein-based material will be realized. Surface charged protein containers can be combined with inorganic compounds to unite biological features with the chemical and physical properties of abiotic materials. In particular, protein containers, with their inherent ability to encapsulate cargo molecules, are perfect platforms for the generation of multifunctional assemblies.[1, 2] For this container-filling gold nanoparticles can be decorated with a small number of encapsulin cargo-loading peptides. By lock-and-key interaction between the peptides and the peptide-binding pockets on the inner container surface, the nanoparticles will be encapsulated with high efficiency.[3] Crystalline materials are produced, which is crucial for future applications. Because the protein scaffold is independent of the nanoparticle cargo, this modular approach will enable tuning of the optical properties by choice of nanoparticle content, assembly type and protein container type. For further future applications, surface charged protein containers will be used as sustainable building blocks for bioinorganic nanomaterials.[4]

[1] M. Künzle, T. Eckert, T. Beck *J. Am. Chem. Soc.* 2016, *138*, 12731.
[2] M. Lach, M. Künzle, T. Beck *Chem. Eur. J.* 2017, *23*, 17482.
[3] M. Künzle, J. Mangler, M. Lach, T. Beck *Nanoscale* 2018, *10*, 22917-22926.
[4] M. Lach, M. Künzle, T. Beck, *Biochemistry* 2019, *58*, 140-141.

P101

MetalJet source enabling advanced protein crystallography

E. Espes[1], J. Hållstedt[1]
[1]Excillum AB, Kista, Sweden

X-ray diffraction data from protein and bio crystals are typically of a much lower resolution and quality compared to measurements in chemical crystallography. This is mainly due to large molecules, poor crystalline quality in combination with very light atoms embedded in large amount of water. At the same time, unit cell size is extremely large and thus the Bragg reflections are very closely packed together.

Protein crystallographers must rely on the strongest X-ray sources to combat the issues of air sensitivity, small crystals, weakly diffracting and densely packed reflections. Traditionally, a high brilliance synchrotron has been used to measure full protein data leading for structure determination, whilst home laboratory instruments have been used only for protein screening to identify the preferred crystals for measurement at the synchrotron. High brightness X-ray compact sources, such as the MetalJet have not only improved crystal screening but also made it possible to obtain publishable data from extremely difficult experiments such as GPCR membrane proteins.

In this communication we will show a number of different examples how the MetalJet X-ray source can be a game changer for home laboratory macro molecular diffraction.

P102

Encapsulation of inorganic nanoparticles into protein containers towards highly ordered biohybrid material

M. Rütten[1], T. Beck[1]
[1]University of Hamburg, Institute of Physical Chemistry, Hamburg, Germany

A synthetic strategy towards optical materials is established using protein containers and nanoparticles. With protein containers as

P103

SAD phasing using a novel gadolinium complex

D. Gilzer[1], L. Venne[1], T. Glaser[1], H. H. Niemann[1]
[1]Bielefeld University, Department of Chemistry, Bielefeld, Germany

Gadolinium complexes have been employed to solve protein structures (Girard *et al.*, 2002) and were commercialized (Stelter *et al.*, 2014) due to their high f''-values at Cu $K\alpha$ wavelength. We employed Gd-julia, a novel Gd-complex, for phase determination of two model proteins. Gd-julia is chemically similar to Gd-DOTA (Fig. 1), a contrasting agent in magnetic resonance imaging that has been established as Gd carrier for crystal soaking (Girard *et al.*, 2002).

Hen egg-white lysozyme (HEWL) and thaumatin crystals were soaked in mother liquor containing 100 mM Gd-julia for 60 minutes before freezing. The crystals showed no damage and produced data to a resolution <2 Å. Data were collected for approximately 1 hour and 27 hours for HEWL and thaumatin, respectively, using a Cu sealed tube microfocus X-ray source with CCD detector and were processed with *XDS*.

Substructure determination in *ShelxD* was straight-forward and automated model building in *PHENIX AutoBuild* produced a nearly complete model. In both cases, the final map showed clear electron density only for the Gd atom of Gd-julia (Fig. 2). Given the distance to the nearest protein side chains, we argue that Gd remains bound to its ligand, which binds to a specific location in the crystal but appears to be rotationally disordered.

To validate that phasing was facilitated by the bound Gd, we prepared native HEWL and thaumatin crystals and recorded data to similar resolution and multiplicity on the same diffractometer. Despite the strong anomalous signal from S and Cl atoms in the native data, substructure determination failed in these negative controls.

Currently, we are extending soaking experiments to other protein crystals.

References

Girard, É. (2003), *Acta Cryst.* D59, 118-126.
Stelter, M. (2014), *Acta Cryst.* D70, 1506-1516.

Fig. 1: Comparison of DOTA and H_4julia ligands.

Fig. 2: $2F_O$-F_C (1σ) and anomalous difference (3σ) density of the Gd-julia binding site in thaumatin.

Figure 1

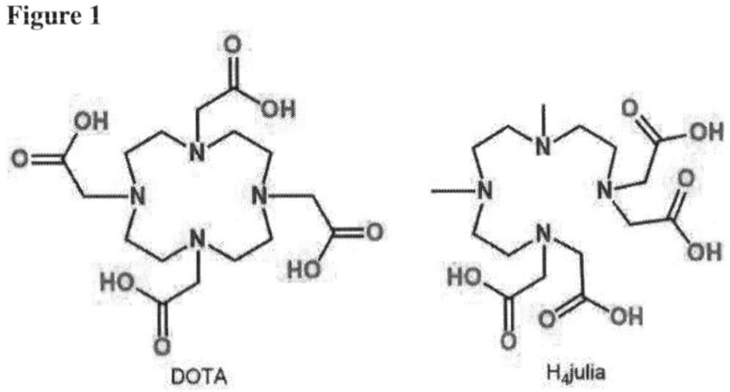

Figure 2

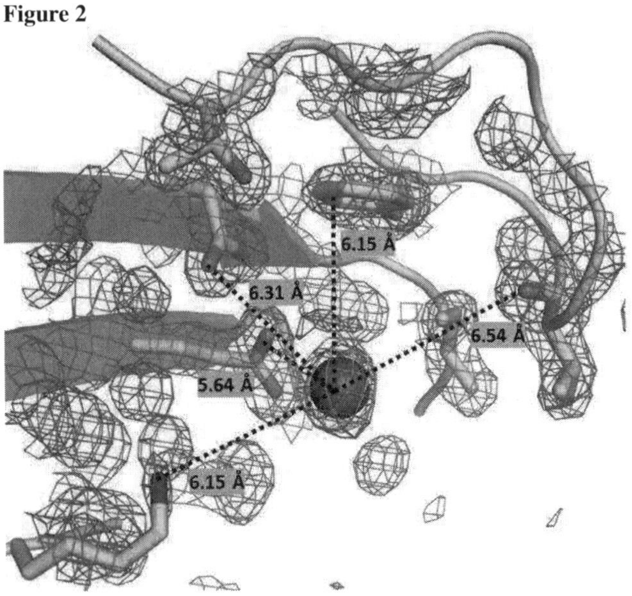

P104

In-house PX Systems - Bone or Bane?

M. Adam[1], M. Mrosek[1], J. Luebben[1], V. Smith[1], H. Ott[1]
[1]Bruker AXS GmbH, Karlsruhe, Germany

The number of structure deposited in the PDB has impressively increased by almost a factor of 20 over the last two decades. Despite complimentary methods having gained increased visibility over the last couple of years, the majority of structures is still determined by classical X-ray methods. This increase of structures goes along with better access to synchrotrons and their high end PX beamlines. Many of these beamlines offer rapid, automated and remote data acquisition services, which were expected to decrease the need for in-house X-ray facilities since also crystal screening experiments could be done at synchrotrons more efficiently than on the home source.

However, the development of high-end components for in-house system with X-ray sources such as the METALJET D2 PLUS or the IμS DIAMOND together with large active-area, photon-counting detectors, such as the PHOTON III M28 have changed that picture. Today"s in-house systems provide impressive data quality and throughput. These systems are completed by user friendly *in-situ* screening units, which can be mounted and unmounted in almost no time. Large research groups benefit from automated cryogenic sample changers, which have been developed to ensure throughput and reliability with a focus on perfect sample recovery.

We will report on latest achievements in the development of in-house systems, cryogenic sample handling systems and their potential benefit for a more efficient preparation of synchrotron trips as well as increased productivity in the home lab.

Figure 1

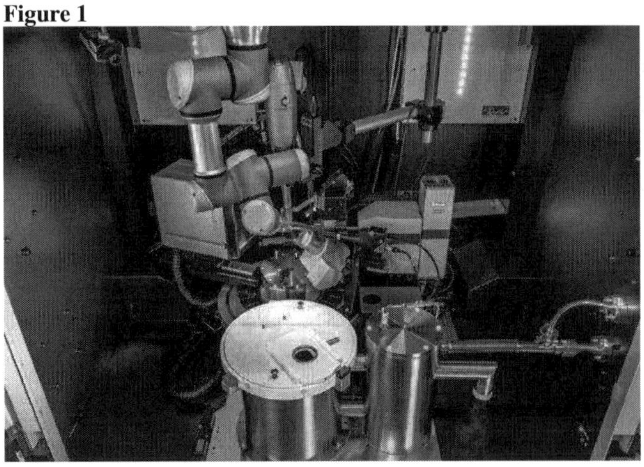

Extreme/non-ambient conditions

P105

What can we learn from high pressure protein crystallography?

K. Kurpiewska[1], K. Lewiński[1]
[1]Jagiellonian University, Faculty of Chemistry, Department of Crystal Chemistry and Crystal Physics , Kraków, Poland

Question

The field of high pressure protein crystallography (HPPX) is now reaching towards its potential to becoming an integral part of biological science with applications in: studies of the high-energy conformers of proteins, investigation of interactions between molecules, studies of the structural principles of organization of oligomeric proteins, pressure induced disorder-order transition in the crystalline state, structure-based prediction of various thermodynamic parameters, studies of the mechanisms of enzymatic reactions and studies of proteins from piezophiles. Are there any new lessons from the field?

Methods

A macromolecular crystal is sufficiently plastic to accommodate relatively large conformational changes. Historically, crystallographic research on the influence of high pressure on protein structures was initiated in 1987 by Kundrot and Richards. Since then number of studies regarding high pressure protein crystallography proved the potential importance of this method for current structural biology. The particular experimental conditions refer to high pressure device (diamond anvil cell) shown at Figure 1, monitoring pressure and viewing sample, sample preparation, data collection and processing. All steps can be implemented in laboratory equiped with standard XRD instrument.

Results

The influence of pressure on proteins, technical aspects of high pressure diffraction data collection and achievements for different macromolecules (ribunuclease A, β-lactoblobulin and insulin) investigated under non-ambient conditions are presented.

Conclusions

The remarkable adaptation of macromolecular crystals to high pressure perturbation can be exploited for the investigation of important aspects of pressure's influence on macromolecules, including conformational changes, dissociation, flexibility. HPPX can also addres other properties crucial for biological function of proteins for instance epitope rearangements and pre-aggregation states.

Figure 1

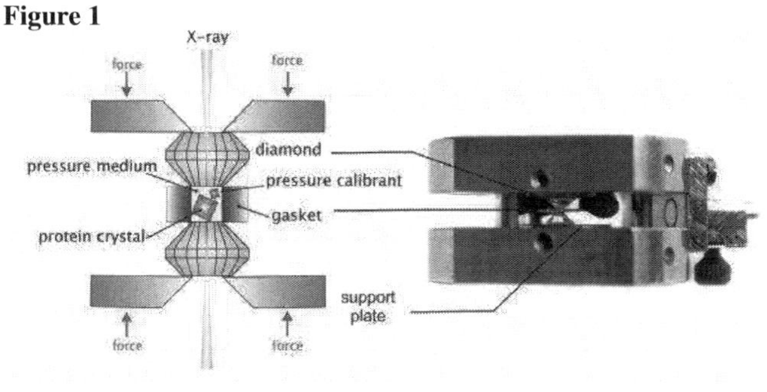

P107

Pressure-induced disulfide methathesis reactions

P. Ratajczyk[1], S. Sobczak[1], A. Katrusiak[1]
[1]Adam Mickiewicz University , Department of Materials Chemistry , Poznań, Poland

The search for new, environmentally friendly chemical reactions is becoming a growing challenge for scientists from around the world. A novel approach to solving this problem is conducting a chemical reaction under high-pressure conditions. The use of high-pressure can lead to the discovery of new materials with unique properties, however, this area still remains unexplored.

In our experiments, we have studied dissociation and formation of a covalent bond during the pressure-induced metathesis of aryl disulphides. This reaction is the first example of a reaction in which the high-pressure is used instead of a catalytic or reducing agent. At ambient conditions, disulphides metathesis requires long equilibrium times and a combination of a reducing agent and base, which limits a practical application of this reaction. Up to this point, only two alternative approaches for facilitating disulphide metathesis have been discovered, however, only the use of the high-pressure allowed for the complete conversion of substrates to products.[1]

Here we further explore reactions between disulphides in order to understand the nature and mechanism of pressure-induced metathesis. For that purpose homodimeric disulphides substituted with either electron-donating group (like $-CH_3$, $-NH_2$) or electron-withdrawing group (-COOH, $-NO_2$, -Cl, -Br, -F) were chosen. The different techniques as PXRD, SXRD, Raman or mass spectroscopy were used for the monitoring of reaction progress. For each metathesis reaction, the conditions were substantially improved until the heterodimeric products were obtained in the form of high-quality single crystals.

This work was supported by funding from the Polish National Science Centre (Preludium No. 2017/27/N/ST5/00693)

[1] S. Sobczak, W. Drożdż, G. I. Lampronti, A. M. Belenguer, A.Katrusiak, A. R. Stefankiewicz, Chem.: Eur. J. 2018, 24, 8769 8773.

Quantum crystallography

P108
Single crystal growth of Alkali Platinum Group Metal Oxides

L. Kerkhoff[1], P. Becker[1]
[1]Section Crystallography, Institute of Geology and Mineralogy, University of Cologne, Cologne, Germany

Recently, Alkali Platinum Group Metal Oxides attracted considerable attention with exhibiting unconventional magnetism (Li_2RuO_3;Li_3RuO_4;Li_2IrO_3) [1-3] and charge ordering phenomena ($Na_{3-x}Ru_4O_9$) [4].

The main structural features of the investigated compounds are their [MO_6]-octahedra (M=Ir;Ru). In the three modifications of Li_2IrO_3 ((α-,β- and γ-Li_2IrO_3 [5-7]) and Li_2RuO_3 [8] edge-shared [MO_6]-octahedra build up a honeycomb-like network. In Li_3RuO_4, edge-shared [RuO_6]-octahedra form zigzag-chains [9], whereas in $Na_{3-x}Ru_4O_9$ manifold chains build up a tunnel-type structure [10].

The need for high-quality single crystals for further investigations turns all compounds into subjects of crystal growth efforts, which are complemented by DTA and both single crystal- and powder-XRD.

Single crystals are grown from carbonatic melt as well as from the gaseous phase. For the latter, a modified setup after [11] was realised, in which educts are separated by spacers with spikes, acting as preferred nucleation sites. The place of crystallisation depends on the position of the chemical equilibrium. Moreover, preferred growth conditions are highly dependent on the stoichiometry of the compounds (lithium ruthenates) and on the structural modification (Li_2IrO_3).

Currently, single crystals of α-Li_2IrO_3(1mm), β-Li_2IrO_3(0.5mm), γ-Li_2IrO_3(0.2mm), Li_2RuO_3(0.6mm), Li_3RuO_4(0.2mm) and $Na_{3-x}Ru_4O_9$(<50µm) can be grown successfully.

[1]Jackeli & Khaliullin,*Phys.Rev.Lett.,*102,017205(2009)
[2]Chaloupka et al.,*Phys.Rev.Lett.,*105,027204(2010).
[3]Manuel et al.,*Phys.Rev.B,*84,174430(2011).
[4]Yogi et al.,*Phys.Rev.B.*,98,085113(2018).
[5]O"Malley et al.,*J.Solid State Chem.,*181,1803–1809(2008).
[6]Biffin et al.,*Phys.Rev.B*,90,205116(2014).
[7]Modic et al.,*Nat.Commun.*,5,4203(2014).
[8]Kobayashi et al.,*Solid State Ionics*,82,25-31(1995).
[9]Alexander et al.,*J.Mater.Chem.,*13,2612-2616(2003).
[10]Regan et al.,*J.Solid State Chem.,*179,195–204(2006).
[11]Freund et al.,*Scientific reports*,6,35362(2016).

Disordered Materials and Complex and aperiodic structures

P109

Ion-irradiation induced swelling and structural changes of diamonds

K. Bunk[1], W. Morgenroth [1], I. Alencar[2], M. Hanefeld[3], C. Trautmann[4,5], B. Winkler[1]
[1]Goethe University Frankfurt, Mineralogy , Frankfurt/Main, Germany
[2]Federal University of Rio Grande do Sul , Rio Grande do Sul, Brazil
[3]Goethe University Frankfurt, Physics, Frankfurt/Main, Germany
[4]TU Darmstadt , Material Science , Darmstadt, Germany
[5]GSI, Material Science, Darmstadt, Germany

Diamond is used in high-dose environments as window material or in radiation detectors [1, 2]. For these applications, a radiation hard material is required. Radiation induces a plethora of effects, which have extensively been studied, but there are still numerous questions concerning irradiation-induced changes in the structure and properties of diamond. Ion irradiation experiments were carried out on synthetic diamond samples, including single crystal, polycrystalline and boron doped samples of type Ia, IIa, Ib, IIb. The samples were irradiated with 14 MeV Au6+ ions and a maximal fluence of 2.4e15 ions/cm2, and with 1.7 GeV Au+ ions with a fluence up to 1e13 ions/cm2. The samples and the induced radiation damage were characterized by Raman spectroscopy, atomic force microscopy (AFM) and scanning electron microscopy (SEM) studies.

AFM measurements show a fluence dependent swelling of diamond due to ion implantation. The corresponding density decrease vs. fluence is shown in Fig. 1. Raman spectra indicate the appearance of sp2-hybridized carbon in the irradiated volume (Fig. 2).

Funding by GSI, BMBF (05K16RFA, 05K16RFB) and Willkomm-Stiftung is gratefully acknowledged.

Figure 1

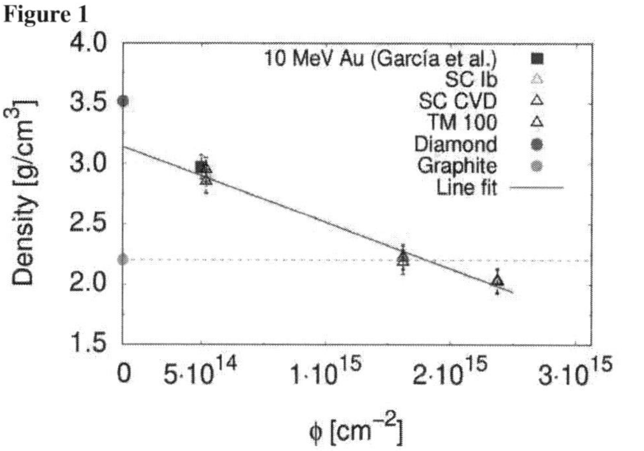

Figure 2

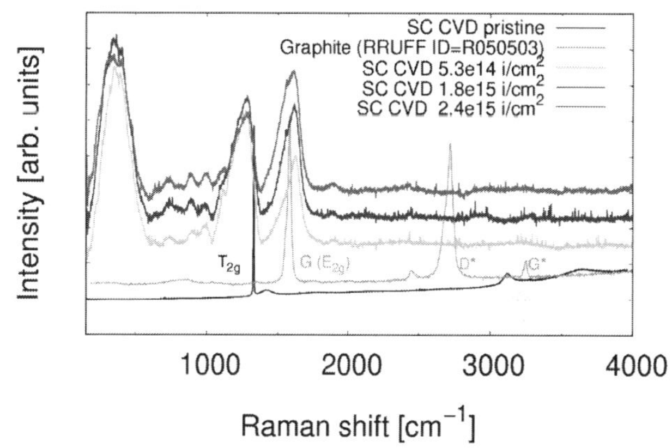

Fig. 1: Density vs. fluence for single crystal (SC CVD, SC Ib) and polycrystalline (TM 100) samples irradiated with 14 MeV Au ions (fluence up to 2.4e15 ions/cm2) and with linear extrapolation of the data. Data from Garcia et al. [3] with 10 MeV Au ions (5e14 ions/cm2) are shown for comparison.

Fig. 2: Raman spectra for pristine and 14 MeV Au ion-irradiated (fluence up to 2.4e15 ions/cm2) single crystal diamond and graphite [4].

[1] Wallny et al., Nucl. Instrum. Methods Phys. Res A, 582, 824-828 (2007)
[2] W. Schildkamp and L. Nikitina, Review of Scientific Instruments, 83, 095104, (2012)
[3] García et al., Diamond and Related Materials, 69, 1-7, (2016)
[4] http://rruff.info/graphite/display=default/R050503

P110

MEAD and cation distribution in CZTSe, CFTS, and CZSiSe

D. M. Többens[1], G. Gurieva[1], S. Niedenzu[1], G. Schuck[1], I. Zizak[1], S. Schorr[1]
[1]Helmholtz-Zentrum Berlin für Materialien und Energie GmbH, Structure and Dynamics of Energy Materials, Berlin, Germany

Multiple Edge Anomalous Diffraction (MEAD) [1] has been applied to various quaternary metal sulfides belonging to the adamantine compound family in order to validate the distribution of copper, zink, and iron cations in the structure. Semiconductors from this group of materials are promising candidates for photovoltaic applications. Their properties are strongly depended on point defects, in particular related to cation order-disorder. However, Cu^{+1}, Zn^{2+}, and Fe^{2+} have very similar scattering factors and are all but undistinguishable by normal X-ray diffraction. Anomalous diffraction utilizes the dependency of the atomic scattering factors *f*' and *f*'' from the energy of the radiation, especially close to the element-specific absorption edges. In the technique called MEAD individual Bragg peaks are tracked over an absorption edge. The intensity changes depending on the structure factor can be highly characteristic for Miller indices selected for a specific structural problem, but require very exact measurements. Beamline KMC-2 [2] at synchrotron BESSY II, Berlin has been recently upgraded for this technique. We applied the method to $Cu_2ZnSnSe_4$, Cu_2FeSnS_4, and $Cu_2ZnSiSe_4$. In spite of all samples being well characterized, having been used in previous studies, MEAD allowed the determination of surprising structural features. Anomalous X-ray powder diffraction and EXAFS compliment the data.

[1] B. A. Collins *et al.*: *Physical Review B* **92**, 224108 (2015)
[2] Helmholtz-Zentrum Berlin für Materialien und Energie: *J large-scale research facilities*, **2**, A49 (2016)

P111

Synthesis and structural characterization of new dabco-templated metal sulfates

T. Bednarchuk[1]
[1]Institute of Low Temperature and Structure Research, Polish Academy of Sciences, Division of Structural Research, Wrocław, Poland

The research on the organic-inorganic hybrid compounds which often exhibit structural phase transitions still creates a significant amount of interest in the scientific community. As a result of the detailed study of new solid-state materials with Sulfur-Oxygen-Metal linkages, four new metal sulfates templated by mono-$(HDABCO)^+$ or di-protonated $(H_2DABCO)^{2+}$ cations were synthesized.

The first compound $(C_6H_{13}N_2)_2[Co(H_2O)_6](SO_4)_2 \cdot 3H_2O$ (I) crystallizes in *P*-1 space group with Z = 2. The independent part of the unit cell is composed of two $(HDABCO)^+$ cations, two SO_4^{2-} anions, two halves of two $Co(H_2O)_6$ octahedrons (lies on the center of inversion), and three solvent water molecules.

The structure of compound $[Cd(H_2O)_4(C_6H_{13}N_2)_2](SO_4)_2 \cdot 2H_2O$ (II) was solved in the orthorhombic *Pbca* space group with Z = 4. The molecular structure consists of one half of a cationic entity $[Cd(H_2O)_4(C_6H_{13}N_2)_2]^{4+}$, one sulfate anions and one lattice water molecule.

The compound $(C_6H_{14}N_2)[Zn(H_2O)_6](SO_4)_2$ (III) crystallizes in the monoclinic $P2_1/c$ space group. The asymmetric part of the unit cell contains one $(H_2DABCO)^{2+}$ cation, one hexaaqua-coordinated Zn cation, and two sulfate anions. In comparison with compounds (I)-(II), the compound (III) is characterized by the presence of di-protonated $(H_2DABCO)^{2+}$ cations and the absence of solvation water molecules. Furthermore, on cooling it undergoes an iso-symmetric structural phase transition. The structural details in both phases will be analyzed.

The structures of compound $[Ni(H_2O)_4(C_6H_{13}N_2)_2](SO_4)_2 \cdot 4H_2O$ (IV) were determined at RT (295 K) and LT (100 K), due to the presence of the temperature-dependent phase transition from monoclinic *C2/m* to another monoclinic $P2_1/c$ space group. The RT structure is characterized by highly disordered cationic and anionic entities, which are fully ordered in the low temperatures.

The results of studies on the structures of new compounds investigated using the X-ray diffraction technique will be presented.

P112

The crystal chemistry of fedorite from Murun massif and its relationship to the structure of martinite

E. Kaneva[1,2]
[1]Saint Petersburg State University, Institute of Earth Science, Saint Petersburg, Russian Federation
[2]AP Vinogradov Institute of Geochemistry SB RAS, X-ray methods of analysis, Irkutsk, Russian Federation

Fedorite and martinite are the rare phyllosilicates. Fedorite, (Na,K)2-3(Ca4Na3)(Si,Al)16O38(F,Cl)2·3.5H2O, occurs on Turii and Malyy Murun alkaline massifs (Russia) [1]. Martinite, (Na,□,Ca)12Ca4(Si,S,B)14B2O38(OH,Cl)2F2·3.5H2O, was discovered at the only Mt Saint-Hilaire deposit (Canada) [2]. The largest open database [3] indicates that the minerals are isostructural. Their crystal structures consist of (Ca,Na)-octahedral layers bonded to tetrahedral (T) layers. Between the two T-layers there are H2O molecules and Na (and K) atoms.

In this work a re-appraisal of Murun fedorite structure was undertaken and its relationship to martinite was reviewed. The structure has been refined in sp. gr. $P\bar{1}$ (R = 2.98), using the following lattice parameters: a = 9.6560(7), b = 9.6633(8), c = 12.644(1) Å, α = 102.534(3), β = 96.215(3), γ = 119.931(2) °, V = 964.79(8) Å3, a/b = 0.9992, b/c = 0.7643, c/a = 1.3094. The structural formula of fedorite sample reveals as: (Na1.16K0.90)(Ca4.08Na3.21)Si16O38F2·4H2O. The refinement shows that *A2* site is spitted in three different positions, occupied by K and Na. An additional position of H2O and a greater water content with respect to [1] are noted.

Martinite (sp. gr. $P\bar{1}$) has similar parameters (a = 9.5437(7), b = 9.5349(6), c = 14.027(1) Å, α = 108.943(1), β = 74.154(1), γ = 119.780(1) °, V = 1038.1(1) Å3, a/b = 1.0009, b/c = 0.6798, c/a = 1.4697 [2]), but its c parameter is slightly greater. The reason is that two symmetrically related T-layers (T16O38) are condensed in fedorite, whereas in martinite two different T-layers (T8O20) are not interconnected.

Thus, the minerals are not isostructural, as fedorite is a silicate with double T-layers while martinite contains isolated T-layers.

This work was supported by grant of the President of the Russian Federation MK-936.2019.5.

References

[1] Mitchell R.H., Burns P.C. Can. Min. (2001), 39, 769 – 777.
[2] McDonald A.M., Chao G.Y. Can. Min. (2007), 45, 1281 – 1292.
[3] Mindat.org

P113

First order CDW phase transition in $Er_2Ir_3Si_5$

S. van Smaalen[1], S. Ramakrishnan[1], A. Schönleber[1], T. Rekis[1], N. van Well[1], L. Noohinejad[1], B. Bag[2], A. Thamizhavel[2], S. Ramakrishnan[2], D. Pal[2], M. Tolkiehn[3], C. Paulmann[4]
[1]University of Bayreuth, Laboratory of Crystallography, Bayreuth, Germany
[2]Tata Institute of Fundamental Research, Department of Condensed Matter Physics and Materials Science, Mumbai, India
[3]Deutsches Elektronen Synchrotron (DESY), Beamline P24, PETRA III, Hamburg, Germany
[4]University of Hamburg, Mineralogisch-Petrographisches Institut, Hamburg, Germany

$Er_2Ir_3Si_5$ at room temperature crystallizes in the orthorhombic $U_2Co_3Si_5$ structure type [1]. The compound was reported to undergo a first-order charge-density-wave (CDW) transition below 166 K based on resistivity (ρ) and magnetic susceptibility (χ) measurements [1]. Similar compounds like $Lu_2Ir_3Si_5$ where investigated through X-ray powder diffraction, stating that the symmetry is lowered from orthorhombic Ibam to monoclinic $P2_1/m$ at low temperatures [2]. $Lu_2Ir_3Si_5$ was also studied through TEM experiments, where the wave vector was reported to be q = δ (-1, 2, 1) where δ ≈ 0.23 [3]. We have studied $Er_2Ir_3Si_5$ via single-crystal X-ray diffraction at temperatures down to 20 K. We report the incommensurately modulated crystal structure in relation to the physical properties of $Er_2Ir_3Si_5$, elucidating the unique nature of the CDW.

Acknowledgement: Single-crystal X-ray diffraction data were collected at Beamline P24 of PETRA-III at DESY, Hamburg, Germany.

References

[1] P. C. Lalngilneia, A. Thamizhavel, S. Ramakrishnan and D. Pal, AIP Conf. Proceed. 1591, 113 (2014).
[2] Y. Singh, D. Pal, S. Ramakrishnan, A. M. Awasthi and S. K. Malik, Phys. Rev. B 71, 045109 (2005).
[3] M. H. Lee, C. H. Chen, M. -W. Chu, C. S. Lue, and Y. K. Kuo, Phys. Rev. B 83, 155121 (2011).

P114

Phase transitions and structural development of FeOCl at high-pressure low-temperature conditions

A. M. Schaller[1], M. Bykov[2,3], E. Bykova[2], K. Glazyrin[4], S. van Smaalen[1]
[1]University of Bayreuth, Laboratory of Crystallography, Bayreuth, Germany
[2]Carnegie Institution of Washington, Geophysical Laboratory, Washington DC, United States
[3]Howard University, Washington DC, United States
[4]Deutsches Elektronen-Synchrotron (DESY), Photon Sciences, Hamburg, Germany

FeOCl belongs to a class of low-dimensional magnetic compounds MOCl. Their dimensionality depends on the orbital order of the 3*d* electrons. For TiOCl that results in a quasi-1D magnetic system with a spin Peierls state at low temperatures [1]. M = V, Cr, Fe form quasi-2D magnetic systems with antiferromagnetic order and strong magneto-elastic coupling at low temperatures [2]. A structural normal-to-incommensurate transition occurs for all MOCl compounds at ≈ 15 GPa, which is not associated with changes in the magnetic order or electronic structure [3]. The application of hydrostatic pressure allows us to continuously adjust the intra- and interchain magnetic exchange parameters vía changing the geometry of the M^{3+} surroundings and M ··· M distances. Pressurizing MOCl compounds above and below the Néel temperature T_N can therefore be used to investigate and to disentangle the complex interplay between structural changes and magnetic order. We carried out high-pressure (HP) low-temperature (LT) single crystal X-ray diffraction experiments on FeOCl at P02.2/PETRA III above and below $T_{N,1\ bar}$ up to 38 GPa. The experiments revealed the pressures of the magnetic transitions, the (super)space groups and crystal structures and provided an insight into the HP-LT mechanisms: at $T < T_N$ the magneto-elastic coupling is dominated by a monoclinic lattice distortion, whereas the magneto-elastic coupling at $T > T_N$ is characterized by an interplay between lattice distortion and structural changes that significantly enhances, from a geometrical point of view, superexchange interactions. Understanding those mechanisms with varying external parameters (P, T, composition) provides crucial information for MOCl compounds and for low-dimensional magnetic systems in general.

[1] Seidel et al., Phys. Rev. B 67, 020405(R) (2003).
[2] Zhang et al., Phys. Rev. B 86, 134428 (2012).
[3] Bykov et al., Sci. Rep. 5, 9647 (2015).

P115

Ferroelectricity in modulated supramolecular chains of phenazinebromanilic acid.

L. Noohinejad[1], M. Tolkiehn[1], M. Kamiński[1], S. van Smaalen[2]
[1]Deutsches Elektronen Synchrotron (DESY), Photon Sciences, Hamburg, Germany
[2]University of Bayreuth, Laboratory of Crystallography, Bayreuth, Germany

In continuation of our interest in aperiodic molecular ferroelectrics [1-2], we present a study of the ferroelectric phase transition in the co-crystal phenazinebromanilic acid. We report temperature-dependent single crystal X-ray diffraction and relative crystal structures at modulated phases. Our additional interest originates from the mobile aspect of hydrogen atoms, which often controls the transformation of the solid-state crystal structures and their symmetries. Temperature dependence of relative dielectric constants and spontaneous polarization of phenazine-bromanilic acid measured along the crystallographic b-axis showed ferroelectricity at lower temperatures [3]. The single crystal data were collected by using the synchrotron radiation at P24 beamline, PETRA III, DESY, Hamburg. The co-crystal is paraelectric at room temperature with centrosymmetric space group P21/n, by lowering the temperature it undergoes two ferroelectric phase transitions. Below 138 K the symmetry breaks to non-centrosymmetric polar space group P21 further lowering the temperature to 101 K it passes through an incommensurate ferroelectric phase where it is finally locked in at 98 K. The results provide the temperature dependence of the charge transfer in hydrogen bonds, including the temperature dependence of the contributions of the various atoms to the ferroelectricity.

[1] L. Noohinejad, S. Mondal, A. Wölfel, A. Schönleber, S. van Smaalen, J. Chem. Cryst. 44, 8, 387 (2014).
[2] L. Noohinejad, S. Mondal, SK. Ali, S. Dey, S. van Smaalen, A. Schoenleber, Acta. Cryst. B. 71,228 (2015).
[3] S. Horiuchi, R. Kumai, Y. Tokura, J. Mater. Chem., 19, 4421 (2009).

P116

Envelope function analysis of quasicrystals

R. Strzałka[1], J. Wolny[1]
[1]AGH University of Science and Technology, Faculty of Physics and Applied Computer Science, Kraków, Poland

Phase retrieval is a long-term problem in diffraction and structure analysis of not only aperiodic, but also periodic crystals. Over years, many numerical methods were developed to retrieve phases of the structure factor for quasicrystals, including low-density-elimination method [1] or charge flipping algorithm [2,3], implemented in available software for structure solution [4,5]. We present a novel technique, not requiring the iterative Fourier transformations, as an interesting alternative to the problem of phase retrieval.

It is known that the diffraction pattern of quasicrystals can be considered as a periodic series of peaks grouped within envelopes, which are distributed periodically in reciprocal space [6,7]. The period of envelopes occurrence is incommensurate with the period of peaks within the envelopes. The situation is similar to that observed in diffraction patterns of commensurately/incommensurately modulated crystals, with the incommensurateness factor being a golden mean τ (for most of the known quasicrystals). If additionally the centrosymmetricity of the diffraction image is observed, it can be shown that phases of the structure factor of a quasicrystal are 0 or π. Whenever the envelope function reaches (crosses) zero, the phase of all peaks grouped within this envelope changes. Assuming this, the phase retrieval procedure can be moved to the reciprocal space, and phases can be obtained directly from the diffraction patterns. The method was promisingly applied to AlNiCo decagonal quasicrystal in [8]. It does not require the iterative methods described before. Its promising use will be discussed in the presentation and tested against real decagonal system of AlCuRh quasicrystal.

[1] H. Takakura, M. Shiono, T.J. Sato, A. Yamamoto, A.P. Tsai, *Phys. Rev. Lett.* 86 (2001) 236-239.
[2] G. Oszlanyi, A. Suto, *Acta Cryst.A* 60 (2004) 134-141.
[3] L. Palatinus, *Acta Cryst.A* 60 (2004) 604-610.
[4] A. Yamamoto, *Sci. Tech. Adv. Mat.* 9 (2008) 013001.
[5] L. Palatinus, G. Chapuis, *J. Appl. Cryst.*40 (2007) 786-790.
[6] J. Wolny, B. Kozakowski, P. Kuczera, L. Pytlik, R. Strzałka, *Acta Cryst. A* 70 (2014) 181-185.
[7] J. Wolny, P. Kuczera, R. Strzalka, Appl. Phys. Lett. 106 (2015) 131905.
[8] B. Kozakowski, J. Wolny, Aperiodic Crystals, edited by S. Schmid, R. L. Withers & R. Lifshitz, pp. 125-132. Dordrecht, Heidelberg: Springer Science+Business Media (2013).

Instrumentation

P117

Advances in the Parameter Space Concept for Crystal Structure Determination – use of neutron diffraction data and resolution study

M. Zschornak[1,2], M. Nentwich[1], D. C. Meyer[1], A. Kirfel[3], K. Fischer[4]
[1]Technische Universität Bergakademie Freiberg, Institute of Experimental Physics, Freiberg, Germany
[2]TU Chemnitz, Institute of Physics, Chemnitz, Germany
[3]Universität Bonn, Steinmann-Institut für Geologie, Mineralogie und Paläontologie, Bonn, Germany
[4]Universität des Saarlandes, Experimentalphysik, Saarbrücken, Germany

Within the last 15 years, the Parameter Space Concept (PSC) was theoretically developed by Fischer, Kirfel and Zimmermann as alternative approach to solve crystal structures from diffraction intensities without use of Fourier transforms [1-6]. It has already been tested on numerous, partly challenging problems of X-ray diffraction. We here present the applicability for neutron scattering using few degrees of freedom. Central for this approach is the inclusion of negative signs of *f*. As a result, the conventional PSC framework has to be extended, since the definition of the asymmetrical unit in parameter space is no longer unique. In this respect, we discuss new properties of parameter space, in particular symmetries, for first general examples.

Further, we evaluate the resolution limit of the PSC. As an example, a split position of La and Sr with (0, 0, *z*=0.3584) has been investigated in the potential high-temperature superconductor $(La_{0.5}Sr_{1.5})MnO_4$, I4/mmm. A positional shift of the cations in the order of $\Delta z \approx 0.001_5$ (≈ 0.02 Å) has been suggested in literature [7]. Enhancing the scattering difference of La and Sr by f''_{Sr}, a split model of $\Delta z = 0.01_0$ was verified using the PSC within a rather conservative feasibility test [8]. As a result a shift $\Delta z = 0.01_3$ had been determined. We now add to the discussion an evaluation based on two additional model data sets, each with ≤ 10 (00*l*) reflections and varied rel. errors of up to 20%. A graphical representation of the parameter space revealed improvement of resolution with a shift of $\Delta z = 0.01_2 \ldots 0.01_6$ ($\approx 0.1_5 \ldots 0.2_0$ Å).

[1] K. Fischer *et al.*, Z. Krist. 220 643 (2005)
[2] A. Kirfel *et al.*, Z. Krist. 221 673 (2006)
[3] K. Fischer *et al.*, Croat. Chem. Acta 81 381 (2008)
[4] A. Kirfel *et al.*, Z. Krist. 224 325 (2009)
[5] H. Zimmermann *et al.*, Acta Cryst. A65 443 (2009)
[6] A. Kirfel *et al.*, Z. Krist. 225 261 (2010)
[7] T. Lippmann *et al.*, HASYLAB annual reports 583 (2003)
[8] A. Kirfel, K. Fischer, Z. Krist. Suppl. 21 101 (2004)

P118

Value Engineering in Engineering Projects

M. Elshourbagy[1]
[1]El-Mansoura University, Egypt, Public Works, Almansoura, Egypt

Value Engineering (VE) is one tool that can counteract the growing of Engineering Projects problems by providing : cost reduction, process improvement and alternative means and materials for any Engineering Projects .

The main purpose of VE is to deliver the necessary function of a certain facility at the lowest cost.

Table (1) represents the distribution of total costs as expended over the life-cycle of a typical construction project.

Design	**Procurement and Construction**	**Operation and Maintenance**

Table 1 - Life Cycle Cost Distribution

In order to better understand the application of VE, one should first identify the main terms which called Function, cost, worth and value.

The systematic approach of Value Engineering is the job plan. It helps identify high cost areas of the project as well as determine the most economical combination of functions to achieve the task (any Engineering Projects).

Function analysis involves thinking about why an item is necessary, rather than thinking about item itself, that is, its function-oriented rather than item oriented. The function analysis approach leads to more creative solutions than item oriented traditional cost reduction approaches.

P119

Clustering of atom types in University at Buffalo DataBank

P. Rybicka[1], M. Kulik[1], P. M. Dominiak[1]
[1]University of Warsaw, Warsaw, Poland

Independent Atom Model, most widely used in crystal structure refinement, assumes spherical symmetry of the atomic density, ignoring the charge redistribution due to chemical bonding. However, more accurate models of electron density are currently available, such as the multipole model that presents an aspherical approach describing the surroundings of a nucleus far more accurately using the Hansen-Coppens formalism[1]. There are a few databanks of different pseudoatom types and University at Buffalo DataBank (UBDB) is one of the major ones[2]. UBDB contains over 300 different atom types categorized by central element type and chemical surroundings and includes various information about electron density distribution, structure and neighbourhood. In our project, we try to organize these data into clusters based on various factors, such as values of parameters occurring in the Hansen-Coppens equation, symmetry or types and number of neighbours. We extract needed values from databank and transform them into useful datasets and then use Python 3.7 to perform clustering using DBSCAN or K-means algorithms. We take both electron density and topology into account. We observe that DBSCAN performs better than K-means in terms of clustering of the electron density shapes of different atom types and it allows us, for example, to easily discriminate between different elements. This work can be helpful to understand what similarities are observed between given atom types and will make expanding the databank with new atom types much easier.

The authors acknowledge NCN UMO-2017/27/B/ST4/02721 grant.

References

[1] Hansen, N. K., Coppens, P. (1978). Acta Cryst. A34, 909–921.
[2] Kumar, P., Gruza, B., Bojarowski, S. A. & Dominiak, P. M. (2019). Acta Cryst. A75, 398-408.

P120

Challenges of charge density under high pressure – study on α-glycine

S. Sutuła[1], M. Malińska[1], R. Gajda[1], A. Makal[1], M. Stachowicz[2], P. Fertey[3], K. Woźniak[1]
[1]University of Warsaw, Biological and Chemical Research Center, Department of Chemistry, Warsaw, Poland
[2]University of Warsaw, Department of Geology, Warsaw, Poland
[3]Synchrotron SOLEIL, Saint-Aubin, France

As high-pressure experiments are quite common in modern crystallography, it is worth inspecting what kind of information is lost during the measurement with the use of diamond anvil cell (DAC).

Aim of this work is to refine charge density model of α-glycine (P21/n) against data collected at room-temperature under ambient pressure with in-house diffractometer and compare it with the one refined upon data collected under 1.5 GPa with synchrotron radiation in wide-angle DAC.

The collected X-ray data were reduced, merged and used for refinement of the multipole model of electron density in MoPro[1]. The following parameters were refined: positions, ADPs, multipoles, κ parameters, extinction and anharmonic thermal parameters for oxygen atoms. In case of 1.5 GPa-pressure experiment, the raw X-ray diffraction data were converted to typical format extension, reduced and then the same procedure of merging and consecutive steps of refining the model were carried out as it was in the case for the ambient conditions data. Completeness of the 1.5 GPa experiment was 52% to 1.1 Å-1 resolution.

Differences between corresponding structural, thermal and electronic parameters obtained in the above two approaches will we presented in great detail. We have also prepared additional structure refined against X-ray data set where the reflections from the full resolution high-quality experiment were limited to those collected under 1.5 GPa. Comparison of the results of all three refinements allows to estimate the missing structural, thermal and electronic information when incomplete X-ray data sets are used and intensities are corrupted by signals coming from diamonds.

This work was supported by the Foundation for Polish Science, TEAM-TECH Core facility for crystallographic and biophysical research to support the development of medicinal products (co-financed by the European Union under the Regional Development Fund).

[1] Guillot, B. et al., J. Appl. crystallogr., 2001, 34, 214-223.

P121

Pushing the Limits of Microfocus X-Ray Sealed Tube Sources for Biological and Chemical Crystallography

J. Graf[1], T. Stürzer[2], M. Benning[3], H. Ott[2], P. Radcliffe[1], J. Schmidt-May[1], C. Michaelsen[1]
[1]Incoatec GmbH, Geesthacht, Germany
[2]Bruker AXS GmbH, Karlsruhe, Germany
[3]Bruker AXS Inc., Madison, United States

The structure determination on ever smaller and weakly diffracting crystals is one of the biggest challenges in the development of in-house X-ray analytical equipment for chemical and biological crystallography, which continuously raises the requirements for modern X-ray sources and detectors. Over the past 10 years, the performance of both, X-ray sources and detectors, have tremendously improved with the introduction of high-brightness low-power X-ray sources and pixel array detectors.

Nowadays, modern low power microfocus X-ray sealed tube sources, such as the Incoatec Microfocus Soure IμS, define the state-of-the-art for most in-house X-ray diffraction equipment, as they deliver intensities in the range of rotating anodes, yet maintain all the comfort of a sealed tube system.

Throughout the past years, we have continuously improved the performance of the IμS by optimizing critical parameters in the X-ray tube and adapting the X-ray optics, making the IμS the market-leading microfocus sealed tube X-ray source with more than 1000 sources sold to date world-wide. The latest improvement for the IμS 3.0, the first and only microfocus sealed tube source fully optimized for X-ray diffraction applications, is a dedicated multilayer mirror for chemical crystallography which delivers an intensity in the range of $8\cdot 10^{10}$ phts/s/mm^2 with a divergence that matches the typical mosaicity of weakly diffracting chemical samples.

Applications that demand an even higher brightness, such as protein crystallography, benefit from our recently introduced unique new class of microfocus sealed tube sources which uses diamond as a heat sink to cool the anode. This IμS DIAMOND combines the performance of a modern 1 kW microfocus rotating anode with all the comfort of a conventional microfocus sealed tube source, and is now available for Cu-Kα, Mo-Kα and Ag-Kα radiation. We will be presenting selected results showing the impact of these recent developments on the data quality.

P122

The reason of formation of nonattractive S···S contacts in thioamides.

A. Grześkiewicz[1], A. Owczarzak[1], M. Kubicki[1], Z. Dutkiewicz[2], R. Kurczab[3], W. Pietruś[3]
[1]Adam Mickiewicz University, Department of Chemistry, Poznań, Poland
[2]University of Medical Sciences, Department of Chemical Technology of Drugs, Poznań, Poland
[3]Polish Academy of Sciences, Medicinal Chemistry, Kraków, Poland

Introduction

The sulphur·sulphur contacts can be regarded as a homogenous type of chalcogen bond if the electrostatic interaction between σ-hole and the areas of electron density concentration can be observed. Although relatively short contacts between terminal sulphur atoms in the crystal structures of thioamides are quite common, their directionality - understood in this case as pointing to the σ-hole-is low.

Objectives

The overwhelming share of nondirectional S···S contacts (93% of all S···S contacts), together with very short distance between sulphur atoms encourage us to investigate the reason of formation of this "interaction". To do this we selected three compounds, namely, 6-aminothiouracil hydrate (**1**), 2-imidazolidinethione (**2**) and 2-thiazolidinethione (**3**), all with short S···S distances (<3.4Å).

Methods

Experimental charge density studies as well as theoretical calculation were used to investigate the S···S contacts present in selected compounds.

Results

Besides the presence of the bond critical point, the topological analysis of sulphur···sulphur contacts, interaction energy calculations and energy decomposition analysis indicate repulsive character of these contacts. This can be outweighed by the presence of a staple molecule (SM), which is any molecule present in the crystal structure, that interacts with both molecules that form the certain S···S dimer. The Cambridge Structural Database analysis has shown that SM can be found in 77% of all structures with S···S contacts shorter than 3.4 Å and in more than half of the structures with the contacts within the van der Waals radius limit.

Conclusion

The presence of staple molecule(s) enforces the mutual orientation of sulphur atoms in S···S contact . The smaller the distance between sulphur atoms, the more substantial is the presence of staple molecule.

References

A.Owczarzak, Z. Dutkiewicz,R. Kurczab, W.Pietruś, M.Kubicki, A.M. Grześkiewcz, *Cryst. Growth Des.* DOI10.1021/acs.cgd.9b01204

P123

Molecular dynamics investigation of the phenolic oxidative coupling protein Hyp-1

J. Śmietańska[1], T. Kozik[1], R. Strzałka[1], I. Bugański[1], J. Wolny[1], M. Jaskólski[2,3]

[1]AGH University of Science and Technology, Faculty of Physics and Applied Computer Sciences, Kraków, Poland

[2]Institute of Bioorganic Chemistry, Center for Biocrystallographic Research, Poznań, Poland

[3]Adam Mickiewicz University, Department of Chemistry, Poznań, Poland

Up to this date, nearly 85% of protein structures deposited in the RCSB Protein Data Bank (PDB) were solved using X-ray crystallographic methods. However, this technique provides only static, time- and ensemble-average representation of molecular arrangement in crystal. Therefore, calculation of the conformational energy landscape under certain force field connects information about structure and dynamics arising from internal motions of molecules. In our work, we used pico- and nanoscale MD simulation method as energetically-based complementary method for validation and improvement of refined protein structure model. We chose high-resolution crystal structure model of Hyp-1 protein from St. John"s wort [1] with potential in various medicinal therapies and interesting abilities to bind biologically active ligand [2] and form different complex patterns [3]. Verification of average simulated protein structure was conducted by comparison between experimental, refined and calculated from MD model structure factors. Comparison between refined and calculated directly from atomic motions during simulation atomic displacement parameters (ADPs) was performed to check validity of refined thermal factors distribution from Hyp-1 structure. The side-chain angle distributions for different types of amino acids (nonpolar, polar and aromatic) were measured to find energitically preferred rotamer forms, especially in the case of sidechain conformations poorly visible in electron density maps or partially occupied.

[1] K. Michalska, H. Fernandes, M. Sikorski, M. Jaskólski, *J. Struct. Biol.* 169 (2010), 161-171

[2] J.Sliwiak, Z.Dauter, M.Jaskólski, Front. Plant Sci., 7:668.

[3] J. Sliwiak, Z. Dauter, M. Kowiel, A.J. McCoy, R.J. Read, M. Jaskólski, *Acta Cryst.* D71 (2015), 829–843

P124

Neutron Laue diffraction at neutron source BER II

M. Tovar[1]

[1]Helmholtz-Zentrum Berlin für Materialien und Energie GmbH, Berlin, Germany

The FALCON Laue diffractometer at the Berlin neutron source BERII was developed in collaboration with the ILL, Grenoble. It is designed for fast neutron Laue data acquisition and makes use of a white ("pink") neutron beam with wavelength band of 0.8-3.2 Å. Pattern acquisition is performed by means of backscattering and transmission detector consisting of four iCDD cameras each. Orientation analysis of Ca-Nd-Vanadite and Ba-Ni-Phosphate single crystals will be presented. Further examples will illustrate the usage of Laue neutron diffraction to perform 3D grain distribution mapping and indexing of oligocrystalline sampes (Raventos et al., arXiv:1902.03200v1). Due to reactor stop there is ongoing work to transfer the instrument to its new location at the Paul-Scherrer-Institute. Instrument re-commissioning is planned for 2020/21.

Transmission Laue image of $Ca_9Nd(VO_4)_7$

Figure 1

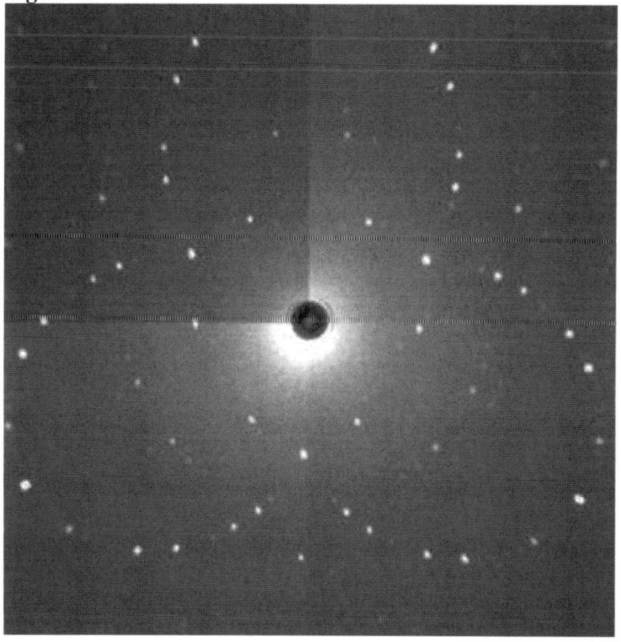

Also of interest

Intermetallics
Synthesis, Structure, Function
Rainer Pöttgen, Dirk Jorendt, 2019
ISBN 978-3-11-063580-5, e-ISBN 978-3-11-063672-7

Crystal Growth of Intermetallics
Peter Gille, Yuri Grin, 2018
ISBN 978-3-11-049584-3, e-ISBN 978-3-11-049678-9

Quantum Crystallography
Fundamentals and Applications
Piero Macchi, 2020
ISBN 978-3-11-060710-9, e-ISBN 978-3-11-060712-3

Series: Zeitschrift für Kristallographie Supplement

Published titles in this series:

Vol. 39: 24th Annual Conference of the German Crystallographic Society, March 25–28, 2019, Leipzig, Germany (2019)

Vol. 38: 26th Annual Conference of the German Crystallographic Society, March 05-08, 2018, Essen, Germany (2018)

Vol. 37: 25th Annual Conference of the German Crystallographic Society, March 27-30, 2017, Karlsruhe, Germany (2017)